Future

Anterior

Future Anterior
Volume IX, Number 1
Summer 2012

Editors' Introduction
Ijlal Muzaffar and Jorge Otero-Pailos

Preservation and Globalization

Across disciplines, the consensus opinion is that the twenty-first century will be defined by globalization. To prepare the professionals of tomorrow, universities are launching initiatives of various sorts and ambition to rethink the form and content of pedagogy.[1] Yet, in most of these enterprises it is presumed that we all share a common understanding and experience of globalization, that we all know what are its characteristics and its future. This issue of *Future Anterior* looks at ways in which problems of preservation help us pause and reexamine what we understand by the term "globalization" and conversely how different ways of thinking about globalization press us to rethink the practices of preservation itself.

Recent scholarship has focused mainly on the role that international heritage institutions have played in advancing globalization. UNESCO's towering position among government-sponsored institutions has made the history of its international preservation campaigns, and its more recent "World Heritage Convention," the subject of important critical scrutiny.[2] Another significant branch of scholarship has focused on the role that heritage institutions play in the commodification of heritage, alongside other private agents (e.g., the global tourism industry) and on the way that development discourse inscribes preservation within a global logic of finance even when it is carried out under the banner of "local" involvement.[3] While in dialogue with this institutionally oriented scholarship, the essays in this issue approach the study of preservation and globalization through different lenses. By shifting away from institutional critique, they discover preservation in unsuspected places, where various understandings and experiences of globalization are in formation and in contention.

Manish Chalana sees preservation emerging in the friction between conflicting class interests. He shows how preservation served as a framework for articulating class-bound notions of the public interest in architecture. The saying that preservation protects the public interest in architecture always refers to a singular public. But Chalana shows there to be multiple conflicting publics in the struggle between the working-class inhabitants of Girangaon, an old industrial district of Mumbai, and new upper-class residents. Architectural preservation has tended to shy away from such debates about gentrification, opting instead to stay narrowly focused on stylistic significance, as a way to

Future Anterior
Volume IX, Number 1
Summer 2012

steer clear of accusations of social engineering. Yet in this case, preservation emerges precisely out of the conflict and out of the desire of working-class residents to shield themselves against the power of newcomers to leverage global capital and to develop what were de facto public parks. Quin Shao also sees preservation as providing a system for expressing the social tensions and differences that arise within competing ideas of globalization. But her study of the Lincoln Lane neighborhood in Shanghai also shows that such celebrations can sometimes be hypocritical, as when so-called preservation experts first defend community preservation plans then turn around and advocate the demolition of those same historic buildings. The presence of tensions between amateurs and professionals is not new, nor is it the exclusive province of preservation,[4] but what is new is the leverage of preservation amateurs in the age of digital social networking and broadcasting. William J. Glover shows how the conflict of values between amateurs and professionals involved in the restoration of Sikh Gurdwaras is not easily resolved, as both make equal claims to represent the interests of global communities whose preservation values are in many ways mutually exclusive. The very nature of the conflict in values makes it nearly impossible to resolve by attempting to incorporate the views of amateurs, who often represent themselves as *the* community, into internationally sanctioned professional preservation practices. This conclusion comes in striking contrast to Paola Demattè's demonstration of the willingness of professional preservationists to reconsider their deontological values in order to accommodate development, especially when it is perceived as inevitable, as in the case of the Three Gorges Dam.

If preservation serves indeed as a system for articulating various experiences of globalization, then the matter of expression, and indeed of preservation's aesthetics, is of central importance. Janike Kampevold Larsen's study of Norway's Tourist Route Project provides a fascinating exploration of one of the central paradoxes of preservation aesthetics, which is that they are aesthetic innovations stemming from efforts to avoid creating new aesthetics. Norway's long tradition of turning landscape painting into preservation efforts is being recast for a global audience of tourists who are invited to bear witness to the similitude between nineteenth-century paintings and contemporary views. For Larsen, the aesthetic expression of sameness enables the production of an entirely new form of intangible heritage, which is as much about the past as about a certain contemporary experience of globalization. Winnie Won Yin Wong shows preservation aesthetics are at the leading edge of a much wider commodification of experience that is a troubling expression of a dominating understanding of globalization. Her critical analysis of the concept of trade dress gives a chilling account of the privatization of areas of experience, including the experience of historic buildings, which were previously unthinkable. However, Janna Israel reminds us that the interweaving of commercial interests and geopolitics is as old as mercantilism and so is preservation's role in giving power aesthetic expression. Her account of the Venetian Republic's attempts to restore the lost Church of the Holy Apostles in Constantinople provides a reminder that some form of globalization has always been afoot but also gives a sense of how these early visions of globalization spurred the creative imagination that went into developing basic preservationist aesthetic forms such as the replica.

Taken together, the essays in this issue signal new and innovative methodological approaches in the study of the relationship between preservation and globalization. Our intention in selecting them was not to seek consensus on what global heritage is. We are mindful that the definition of *heritage* is constantly in flux, as is the definition of *globalization*. We are interested in what these shifts tell us about the place of preservation in particular cultural and political spheres. When the adjective characterizing heritage shifts from *world,* to *international,* to *global,* the institutional mechanisms through which preservation discourse becomes enmeshed with geopolitics become visible on various registers. It is worth recalling, for instance, that the congress that produced the Athens Charter and its famous definition of world heritage was a product of the League of Nations,[5] that "world" heritage reflected a particular view of the world defined by the colonial mandates. The League of Nations' goal of "intellectual co-operation" through conservation projects was to create a diplomatic space that could be used to align the interests of various competing colonial powers, preventing future conflicts. In hindsight, it is easy to see how the change from "world" to "international" cultural heritage under ICOMOS's purview stemmed from how Cold War politics crisscrossed the institutional structure of UNESCO itself.

Yet it would be just as much a mistake to see shifting preservation definitions as simply a *reflection* of geopolitical transformations. These preservation enterprises are mediums through which something called "globalization" is imagined as well. They are the very sites where different ideas of the global and globalization materialize and gain traction. The materiality of preservation projects—their scale, methods of construction, materials and techniques employed—influences what kind of geopolitics can be imagined through them.

This is the view of the interaction between preservation and globalization that is taken by the articles in this issue of

Future Anterior. Examining historical case studies in different geopolitical contexts, historical moments, and institutional domains, each essay examines how in different contexts preservation enterprises shape, interfere, hasten, or slow down the mobilization of political ideas and purported universals. This is a difficult and a critical proposition. Rather than simply faulting each political setting for corrupting some universal idea of preservation or globalization, or blaming the idea itself as exploitative and oppressive, each paper represents how both domains, the material and the discursive, are produced in mutual friction.

This view stands in contrast to what has become the dominant understanding of globalization, and global preservation enterprises, in the Anglo-American academy. The dominant view has come to the surface along the ripple produced by certain major theoretical texts on globalization in the past decade. Michael Hardt and Antonio Negri, Fredric Jameson, Saskia Sassen, Masao Miyoshi, Arjun Appadurai, and Ulf Hannerz are perhaps the most familiar of these authors in academic circles. For these authors, globalization is above all a circuit of flows, a network that has established itself on a global scale and constitutes a new "empire" in the hands of the multifarious agents that ride its currents and extend its reach. This view, however critical, gives globalization a certain mechanical and infrastructural quality. It places it outside the lived human history and imbues its forms with a certain inevitability. It also presents the local and the global as mutually exclusive scales and spheres.

The texts in this issue of *Future Anterior,* however, point toward the increasing presence of another kind of research that challenges this prevalent view of globalization as a self-mobilizing force. For each article, in a unique way, particular enterprises carried out under the banner of globalization aren't examples of something called globalization that is already afoot and bears down on the local from on high; globalization, instead, is a *universal* that is recounted, recalled, and employed to make changes in concrete circumstances. Each time, "globalization" as a universal is invoked differently, for different ends, and produces different results. In this regard, their conceptual thinking lies more closely with authors such as Anna Tsing, Pheng Cheah, James Clifford, and Gayatri Spivak, who have argued that a phenomenon such as globalization cannot be grasped separately from the specific material conditions in which it is made to travel.

This, however, is not to say that universal concepts, or concepts that claim to be universal, do not exist. They very much do. Globalization, preservation, and capitalism are all universals that are invoked by different people in different contexts to "talk across difference," to borrow Anna Tsing's phrase.[6] As Spivak has famously argued, "universal is what we cannot not want," even when it excludes us from its benefits.[7] But paying attention to the "practical" circulations of universals, such as globalization and preservation, in daily encounters changes our understanding of them. The focus becomes on who mobilizes which universals and how. It opens up a space to account for the particular forms that globalization and preservation have taken, locates them in particular biographical, political, and economic contexts, rather than seeing them as a result of some inevitable self-propelling process. It is our hope that this special issue of *Future Anterior* will be able to move the understanding of both the terms in this new direction.

Notes

1 *College Learning for the New Global Century: A Report from the National Leadership Council for Liberal Education and America's Promise* (Washington, D.C.: Association of American Colleges and Universities, 2007).

2 See Jan Turtinen, *UNESCO's World Heritage: On the Preconditions of International Policy* (PhD diss.: Stockholm University, 2006). See also Lucia Allais, *Will to War, Will to Art: Cultural Internationalism and the Modernist Aesthetics of Monuments, 1932–1964* (PhD diss.: Massachusetts Institute of Technology, 2008).

3 See, for instance, *Consuming Tradition, Manufacturing Heritage: Global Norms and Urban Forms in the Age of Tourism,* ed. Nezar AlSayyad (London: Routledge, 2000); Noha Nasser, "Planning for Urban Heritage Places: Reconciling Conservation, Tourism, and Sustainable Development," in *Journal of Planning Literature*, no. 17 (May 2003): 467–79; Rahil Ismail, Brian Shaw, and Ooi Giok Ling, eds., *Southeast Asian Culture and Heritage in a Globalising World: Diverging Identities in a Dynamic Region* (Farnham, U.K.: Ashgate, 2010).

4 Edward Said, "Professionals and Amateurs," in *Representations of the Intellectual*, 65–84 (New York: Knopf, 1996).

5 The conference was organized by the International Museum Office established in 1926 by the International Committee on Intellectual Co-operation of the League of Nations. The Museum Office was part of the International Institute of Intellectual Co-operation at Paris, which was founded in 1922 as an advisory council for the League of Nations. It held such well-known members as Henri Bergson, Albert Einstein, Paul Valéry, Marie Curie, Thomas Mann, Béla Bartók, and Salvador de Madariaga. This membership roster represented a new idea of "intellectual cooperation," that was to provide a counterweight to the competing political and imperial concerns of the League members. This model will serve as a basis for the formation of UNESCO as well, which adopted in 1945 the motto "Since wars begin in the minds of men, it is in the minds of men that the defences of peace must be constructed."

6 Anna Tsing, *An Ethnography of Global Connection* (Princeton, N.J.: Princeton University Press, 2005), 13.

7 Gayatri Spivak, *A Critique of Postcolonial Reason* (Cambridge, Mass.: Harvard University Press, 1999).

In This Issue

Of Mills and Malls: The Future of Urban Industrial Heritage in Neoliberal Mumbai
Manish Chalana

The mandate of historic preservation is to maintain vestiges of diverse cultural heritage, a task that is becoming increasingly difficult in rapidly globalizing India. Much of the country's urban heritage outside of the "monument-and-site" framework is threatened by massive restructuring of cities facilitated by neoliberal urban policies. Mumbai has a rich cultural heritage, associated with diverse sociocultural and economic groups. Much of this is threatened by development practices pursued by various forces with a particular vision of Mumbai as an emerging "global city." In this work Chalana examines Girangaon, an early industrial district of Mumbai, currently being transformed by forces of domestic and global capital. He argues that Girangaon's urban industrial heritage is a significant piece of the city's development history, which future visions of a global metropolis should embrace. While the expansion of Mumbai's economy has benefited some avenues of preservation practice in Mumbai, in Girangaon its consequences have also been negative, as a working-class neighborhood is restructured into a hypermodern district for the elite. The current forms of preservation practice in the city have been insufficient in addressing the complexity around managing heritage in low-income neighborhoods. Girangaon, and Mumbai overall, reveal the many ways that economic, cultural, and political globalization can impact historic preservation practice.

Citizens versus Experts: Historic Preservation in Globalizing Shanghai
Qin Shao

Shao studies citizen preservation in the global city of Shanghai. The pursuit of economic growth has led to large-scale demolition of residential homes and historic structures there since the late 1980s. It has also given rise to citizen involvement in historic preservation. In 2002, Lincoln Lane, a neighborhood built in the 1920s, faced demolition. The residents launched a campaign to preserve their Lane. They documented the significance of the Lane, petitioned the local government, contacted the media in and outside China, and reached out to the people of Springfield, Illinois, President

Future Anterior
Volume IX, Number 1
Summer 2012

Abraham Lincoln's hometown, for support. Some renowned Chinese experts on urban preservation initially supported their causes but eventually sided with the local government, which resulted in the demise of the Lane. This study explores the conflict among the citizens, the experts, and the government, and the politics and economics in historic preservation. It also demonstrates global influence in historic preservation. While the advance of global capitalism was the ultimate cause of the demolition, globalization has also provided the residents with not only a heightened awareness of historic preservation but also more resources to argue for that cause.

Shiny New Buildings: Rebuilding Historic Sikh Gurdwaras in Indian Punjab
William J. Glover
This article examines some of the tensions inherent in conservation and reconstruction work taking place on Sikh religious monuments in northen India, particularly at those monuments that mark important events in Sikh history or in the lives of the Sikh Gurus and their families. Juxtaposing the divergent practices of enthusiastic lay worshippers with those of more professionally trained heritage conservationists, the author argues that both groups engage simultaneously with protocols of objectivist "history" and subjective notions of "heritage," albeit in differing ways. The analysis focuses on a handful of sites undergoing renovation in Anandpur Sahib, India, to explore the material, cultural, and rhetorical strategies through which different groups secure their appropriation of the sites' historical and affective qualities. Two museums located in the town—one dedicated to the life history of Guru Tegh Bahadur, and the other intended to be an internationally prominent center for Sikh heritage—add further complexity to a town landscape replete with meaningful objects.

After the Flood: Cultural Heritage and Cultural Politics in Chongqing Municipality and the Three Gorges Areas, China
Paola Demattè
Following the construction of the Three Gorges Dam and the gradual filling of the lake in recent years, the municipality of Chongqing, which holds most of the massive reservoir, has undergone great ecological and social changes. Aside from environmental problems and social issues, the construction of the dam has inevitably also affected the "local cultural heritage," putting in place new criteria for its definition. Charting the development of preservation debates around the sites affected by the Three Gorges Dam, this essay discusses how conservation and rehabilitation priorities are always tied to cultural politics in various national and international realms.

Global Tourism Practices as Living Heritage: Viewing the Norwegian Tourist Route Project
Janike Kampevold Larsen

This article discusses the Norwegian Tourist Route Project (TRP) in relation to questions of heritage. It frames the TRP as a case of production of heritage, and as a practice that accordingly problematizes the concept of *authenticity* —all in the light of the global tourism industry. It explores the practices that structure the installations in the TRP and suggests that they perform an unwitting form of intangible heritage preservation by encouraging tourists to reenact viewing practices that are deeply founded in a nineteenth-century Norwegian—and European—tradition for viewing landscape. Thus the main dynamic of the project is the construction of space and experience. The TRP reveals a quite random blend of practices, where considerations of local heritage are mostly left to designers' choice and intuition. Still, it taps into historically defined social practices and a globalized design language.

Ambience as Property: Experience, Design, and the Legal Expansion of "Trade Dress"
Winnie Won Yin Wong

Since the 1990s, extensive legal activity has developed around the concept of "trade dress," a subcategory of U.S. trademark law defined as the "total image and overall appearance of a product." This essay examines arguments and claims made for the application of trade dress to designed spaces, from the Taco Cabana fast food chain to Times Square billboards, and tracks an emerging legal definition of ambience or atmosphere as property. Placing this legal development within a broader shift toward experiential concerns in design and preservation practices, the article considers some of the historical questions raised in light of the expansion of legal purview into realms of consumer experience.

A History Built on Ruins: Venice and the Destruction of the Church of the Holy Apostles in Constantinople
Janna Israel

In 1463, construction was underway on the Fatih Mosque in Istanbul by Sultan Mehmet II. According to one of the Sultan's ambassadors, the Mosque was located "in the best site of the city." The site of the new mosque was the Church of the Holy Apostles, which had served as a site of Byzantine imperial ceremony. While the circumstances of its destruction remain murky, Israel argues that the demolition of the church was subsumed into the rhetoric of conflict as Mehmet conquered Venetian territory along the Adriatic, and as Pope Pius II tried to stir enthusiasm for a crusade. Against this backdrop of political strife, this article explores the reproduction and redefinition of the Church of the Holy Apostles through traces of its relics and architectural forms as the Republic again redefined their relationship with the Byzantine Empire, claiming inheritance over its patrimony.

Artist Intervention: The Crystal Life
Roger Hiorns
The chemical changes that building surfaces undergo as they age can have dramatic visual effects. Preservationists scrutinize surface changes to try to establish the fine line that divides those alterations that contribute to the historic significance of the building from those that are detrimental to it. On the basis of this distinction, preservationists determine which changes should be retained and which effaced. Roger Hiorns also manipulates the chemical composition of building surfaces with an eye to how they alter architecture's significance, but his approach irreversibly projects buildings into an alternative almost alchemical reality and temporality. Under his signature coat of copper sulfate crystals, contemporary buildings appear as ancient as unearthed metal tools, encrusted in crystalline forms familiar only to archeologists. His work "Seizure," featured in this issue along with a selection of sculptures, involved covering the entire interior of a brutalist council flat in copper sulfate crystals. It was produced by ArtAngel. "A viscous reality where synthetic overproduction scrapes against the flesh of the visitor. An object that cannot be given a name."

Roger Hiorns is an artist based in London. He was a finalist for the 2009 Turner Prize. His work has been exhibited in major solo shows at museums such as the Tate Britain, The Art Institute of Chicago, and the Aspen Art Museum, among many others.

1. *Chawl* in Lower Parel. Photograph by Sajjad Lambe. Used by permission.

Manish Chalana

Of Mills and Malls

The Future of Urban Industrial Heritage in Neoliberal Mumbai

Mumbai: India's Gateway to the World

India is currently experiencing a powerful wave of cultural and economic globalization. This is certainly not India's first tryst with global exchange; the subcontinent has been a part of various trading networks and empires for millennia. In recent centuries, colonization by the British and other European powers further extended India's global networks through trade and commerce. However, the current wave is unique in its impacts on design and planning. Previous waves of interaction sometimes allowed for a greater degree of indigenization and hybridization, even when, at times, pressure from those in power to subjugate local ideas and understandings of place remained high. This created parallel spaces for various forms of indigenous modernity.[1] However, those agents pursuing the current wave of globalization as a whole seem to devalue such interactions between the local and the global. As a result, cities across the country are allowing their built environments that represent unique cultural preferences and aspirations to be replaced with new ones that reflect the aesthetics and planning ideals of global capitalism with little relation to local context. The consequence of this is that Indian cities are beginning to look increasingly like their Western (and now Far Eastern) counterparts, losing their distinct identities and unique urbanism. These emerging "nonplaces" cater to the needs of a small sliver of the country's elite populations, while simultaneously undermining values and practices inherent in locally evolved built environments.[2]

The case of Mumbai is particularly noteworthy. As the "Gateway of India," the city is the undisputed financial and entertainment capital of the nation. It is one of the largest and densest cities in the world, with 19 million people living in its metropolitan area. The city's built environments include a variety of indigenous and colonial districts and several other hybrid places that interpret colonial modernity in myriad ways. All this gives the city its distinct urban character. However, as a direct consequence of India's economic reforms of the 1990s, Mumbai has begun implementing various neoliberal policies that are reshaping its urban landscapes. Local circumstances cause the implementation of these types of urban policies to intersect with various nationalist, regionalist, and religious politics in unique ways. Although the city has progressive heritage legislation in place, the pressures to redevelop land remain great, due to the

Future Anterior
Volume IX, Number 1
Summer 2012

exceedingly limited supply of buildable land on this peninsular city. The practice of historic preservation is thus pitted against powerful development lobbies whenever it steps outside of its more established "monuments and sites" framework.

The recent expansion of Mumbai's economy through the city's increased role in global markets has surely opened new financial avenues for preservation practice. And the reaction against the rapidity and scale of recent urban changes has led to the emergence of more community engagement with preservation. Both of these phenomena have benefited the field overall. Nevertheless, the current preservation practices in Mumbai continue to emulate methods that are insufficient for low-income, mixed-use context where lifeways, livelihoods and built environments are inextricably linked. These types of neighborhoods also remain the site of conflicting claims about the role of history, and of particular visions of history, within the framework of a "global city." One such area that is particularly threatened is Girangaon—a composite of working-class neighborhoods of mills and worker housing that is being transformed into a hypermodern district of luxurious living and consumption. Located at the very heart of old Mumbai, Girangaon has seen a complex mix of stakeholders struggle for two decades over issues of livelihood, cultural continuity, and memory. Nowhere else in the city more clearly reveals the myriad ways in which forces of globalization intersect with historic preservation over conflicting claims of the city's past and future than in Girangaon.

Girangaon's Industrial Past and Present

Girangaon (or "mill village" in the Marathi language) is a product of a previous wave of economic globalization brought about by British colonialism. It is widely considered the birthplace of India's textile industry, which has been a major driver of the economic, political, and cultural character of Mumbai since the mid-nineteenth century. The Indian cotton boom began with the worldwide shortages of the product created by the American Civil War. The British established Bombay as the "great seat of cotton industry in India" since it possessed, "*par excellence,* all the necessary qualifications for this industry to take deep root and flourish." The labor-intensive industry eventually employed tens of thousands of men, women, and children from as far away as modern-day Pakistan, laying the foundation of the city's multicultural character.[3] Bombay's advantage of cheap labor and raw materials proved to be powerful competition for the English textile industry centered in Lancashire, which lost trade as a result. By the dawn of the twentieth century, the textile industry directly or indirectly supported one-fourth of the city's 800,000 people.[4] Much of this activity was concentrated in a series of districts—Tardeo, Parel, Lalbaug, and Byculla—spread over six hundred acres to the north of the British colonial settlement. These neighborhoods collectively became known as Girangaon.

Girangaon remained an industrial stronghold for more than a century, providing employment to roughly a quarter of the unionized workforce in the city as late as the 1970s. A relatively rapid decline set in after that, brought about by increased competition from powerloom mills, and reluctance on the part of mill owners to upgrade their outdated facilities. As a band-aid measure, the government nationalized several "sick" mills, which delayed the complete demise of the industry. An additional major blow came in 1982 in the form of a prolonged workers' strike. However, the state continued to operate several mills sluggishly (and often at a loss), attempting to revive, and to some extent modernize, the industry.[5] With the neoliberal economic reforms of the early 1990s, the remaining mills all ceased or nearly ceased operations, although worker populations and their descendents have continued to reside in the neighborhoods of Girangaon.

The urban form of Girangaon originally comprised primarily the mill compounds and the neighborhoods of worker housing known as *chawls*. Mill compounds generally occupy ten-to-twelve-acre parcels, within which are located several large-span structures covered with glazed north-light iron trusses.[6] The buildings used basic construction and materials (primarily brick and stone); designs generally reflected the industrial architectural forms from Britain, including the quintessential masonry chimney; however, these were indigenized in a variety of ways using local techniques and materials, such as terracotta tiles as roofing materials. English decorative details accentuated the pediments, cornices, and pilasters. The compound also housed the management quarters and bungalows in gardens containing water bodies and thick foliage.[7]

Outside the mill compounds was the *chawl* neighborhood, where the majority of the workers lived within walking distances of their places of employment. This indigenized barrack-style form possessed all three qualities most desired by the employers and owners: "cheap, convenient, and controllable."[8] It comprised a series of one- or two-room apartments facing an open but covered verandah-hallway, with shared toilets. Although compact, the dwellings accommodated a variety of uses and a host of activities, with the mezzanine and long, open hallways providing additional spaces.[9] The relatively open plan was also well suited for the tropical climate and provided much needed outdoor spaces. As the textile industry expanded, the urban industrial landscape of Girangaon evolved into a dense network of mill compounds, *chawls*, markets and *maidans* (multi-use

2. A typical mill compound in Mumbai. Photograph by Jones Photography. Noncommercial use permitted through the Creative Commons license.

open spaces used for festivities, ceremonies, and play). All these resulted in the emergence of distinct social communities, institutions and patterns of activity in the neighborhood.[10] A century and a half later, the *chawls* have remained active centers of community life and continue to provide affordable housing to large numbers of low-income residents.

Girangaon: Lost in Transition

In the 1990s, India began the long process of liberalizing its economy, a process that has now begun to dramatically reshape its cities. In a radical departure from the quasi-socialist urban policies of the postindependence era, the architects of these reforms pushed for a competitive free market economy with minimum state interference. This has been achieved through a combination of deregulation, privatization, and a dismantling of the welfare state.[11] Mumbai's power brokers have, on the whole, welcomed the reforms, as they allow the city to participate in the global economy more robustly. Several urban policy and vision documents that stem from these reforms impact the transformation of Girangaon directly. Of these, the *Regional Plan for Mumbai Metro Area* (1996–2011) and *Vision Mumbai* (2003) set the stage for these transformations, relying on new urban polices and public–private partnerships to implement them. Both these documents promote a "global business friendly" vision, eying central city locations for primarily office and retail developments.[12] *Vision Mumbai* (prepared by the U.S.-based McKinsey & Company) is the less restrained of the two, as it proposes a Shanghai-like transformation for Mumbai, seeks global investments in billions of dollars, and recommends extensive demolition and rebuilding of old neighborhoods.[13]

Neither proposal would have been possible without a series of regulatory changes building up to them. One of the first of these, and among the most directly relevant to Girangaon, was the repeal of the 1976 Urban Land Ceiling and Regulation Act (ULCRA) in 1999. This act had designated any land over 500 square meters (5400 square feet) owned by a single owner or organization as "surplus," which the state could then claim for development and redistribution.[14] With its repeal, the door was opened for developers to accumulate land holdings for large-scale redevelopment. However, the mills are also subject to the state's Development Control Rules (DCR), given the long-standing commitment within some branches of government to promote and protect the textile industry. Due to pressures of redevelopment, during the early years of economic reforms, the DCR were amended to allow limited nonindustrial developments on the mill lands; however, the revision stipulated that profits must be used in efforts to revive the mills. Several mill owners became creative in developing large portions of their mills for nonindustrial uses in the form of high-end luxury residential, office, and retail uses, keeping only a small fraction of the manufacturing and labor force intact as they outsourced production to suburban venues.[15] Despite this obvious misuse of well-intentioned amendments, it took ten years and prolonged worker and union agitation to amend the DCR in 2001 to introduce provisions for livelihood for jobless workers, tenancy rights in *chawls,* and adequate compensation. It also allowed the division of mill property into three equal parts, to be allocated to affordable housing, civic uses (including open spaces), and market-rate developments. The owners of some properties, however, chose to interpret this tripartite division as applying only to "vacant" (existing nonbuilt) lands on the mill grounds and not to land on which structures existed, and they were supported in this interpretation by the courts. This drastically reduced the proportion of the overall site that an owner would need to allocate for public purposes to about 10 percent or less.[16] As a result, the bulk of mill lands in Girangaon became available for highly profitable market-rate developments, with the support of some state actors, and against the desires of others.

The *chawl* neighborhoods have their own regulatory history. Many of these are in a rather deteriorated state, an unintended consequence of another well-intended policy, the Bombay Rent Control Act. This act mandated extremely low rents and allowed

3. The transforming urban fabric of Girangaon, showing high-rise constructions in the background and High Street Phoenix Shopping Mall—an adaptive reuse of an old mill—in the foreground. Photograph by Amey Kandalgaonkar. Used by permission.

for tenancy to be passed through generations. This dissuaded owners from taking care of their properties, as they saw no benefit in the form of increased rents. As a result, some of the *chawls* have been recognized as "vertical slums," and their redevelopment can be fostered through the Slum Rehabilitation Scheme (SRS) of the Slum Rehabilitation Authority (SRA). At the core of SRS is a public–private partnership model, a hallmark of neoliberal urban policy. It was set up in the 1990s by a coalition led by Shiv Sena, a powerful right-wing Hindu Nationalist party headquartered in the city. In setting up the redevelopment policy, Shiv Sena made claims of moral values, backed by "religion and cultural nationalism,"[17] and presented the scheme as representing "free homes for the poor." Due to its popularity all subsequently elected governments have retained the SRS in some form.

The essence of the SRS program is to provide incentives in the form of increased FSI (floor space index) to developers for rehabilitating the original *chawl* residences in 225 square foot (21 square meter) apartments within the new development. This has spurred a spate of redevelopment activities in both publicly and privately owned *chawls* throughout Girangaon. For example, the 133 historic *chawls* of the Bombay Municipal Corporation (BMC) are being redeveloped into a mini-township.[18] Another example includes three historic *chawls,* spread over eight acres, associated with the Digvijay, Kohinoor, and Tata mills, which the state's National Textile Corporation has already sold for redevelopment, displacing more than 1,200 families.[19]

Nearly all of the current redevelopment projects share common design approaches for reshaping Girangaon. All of these exhibit the ideals of large-scale modernistic planning theory and replicate pervasive Western prototypes exhibiting "hyper-modern" architecture of "non-places."[20] Such alienating urban transformations are a complete departure from the finer-grain urbanity that the mills and *chawls* created. The mill lands are often developed by international architecture and planning companies to portray the image of the city as a familiar place for global capital and commerce. At the same time, despite attempts by some forces within the city to maintain the original community in the neighborhood, little success has been achieved in that regard. On many occasions the original residents are relocated to project-like high-rise buildings in a considerably transformed neighborhood context. The working-class neighborhood is continually being restructured to cater to the lifestyles of the global elite and their interests. With the erasure of the historic urban fabric of Girangaon, the city is rapidly losing virtually all of its physical connection to its own early industrial and cultural history, which allowed it to gain its current prominence.

Preservation Practice in Girangaon: Global and Local Claims

There are indeed forces within the city's preservation community who are well aware of the importance of industrial and working-class history to urban identity; thus, in the midst of all these redevelopments, there are some signs of preservation and adaptive reuse in Girangaon. A portion of the state-operated United Mill has been set aside for a textile museum where the city's industrial history is to be interpreted. The plans are currently under review by the BMC, and after approval, it would take some years for the project to be completed.[21] Some buildings in other mills have found popular reuse as high-end shopping malls. Included in this category is the Phoenix Mill (renamed the High Street Phoenix). Both approaches have been presented as success stories, and they are undoubtedly steps in the right direction. However, their value remains limited. On the one hand, the United Mill represents a classic example of "museumification," with an emphasis on freezing a single piece of the built environment in time, divorced from the lifeways that accompany that environment and were made possible by it. On the other hand, the mall represents an example of adaptive reuse, but the target audience for that use is an economic stratum far above that of the workers whose lives have been intertwined with the mills for generations. Whether those workers continue to live in *chawls,* or in resettlement housing in the immediate vicinity, the malls foster a further divide in the social and cultural life of the community.[22]

Why has there been so little preservation in Girangaon—a site of central importance to the city's historical development? There are multiple pieces to this puzzle. One might imagine a familiar story: preservation practice in India is not integrated with urban planning or is stuck in an old paradigm in which

4. Advertisement for luxury mixed commercial/residential tower in Lower Parel; Hafeez Contractor, architect. Reproduced from *Times of India–Mumbai*, September 17, 2010.

only grand monuments and sites are considered worthy of preservation. Although there are elements of truth to this claim, it is a simplification. For much of the twentieth century, the field in India did remain focused almost entirely on its original mandate of monuments and archaeological sites, as laid out in its various founding acts (the Ancient Monuments and Preservation Act of 1902), agencies (the Archaeological Survey of India), and guidelines (the "Conservation Manual"). Even to this day, the focus of preservation in India remains more concentrated on grand monuments than elsewhere in the world (perhaps partly because India *has* more grand monuments than most other countries). However, there have been several theoretical, practical, and methodological advances to the field in recent decades. Over the years, the definition of heritage has conceptually expanded to include historic zones (similar to the historic districts in the United States), heritage villages, and cultural landscapes. The field has opened up to

5. Composite image of the Blue Frog Lounge and celebrity guests at an event in the same venue in 2011. The Lounge was created through adaptive reuse of a warehouse in the Mathuradas Mill Compound and was designed to maximize the large span and northern light. Photograph by Scott Norsworthy. Used by permission.

the involvement of the private sector and other nongovernmental organizations, including the creation in 1984 of the Indian National Trust for Art and Cultural Heritage (INTACH), a parallel heritage advocacy organization. Recent decades have also seen active participation by world cultural agencies and organizations, particularly UNESCO, as the number of historic properties in India inscribed on the World Heritage list has dramatically increased.[23]

Furthermore, preservation practice in Mumbai is considered to be more progressive than in India generally. For instance, Mumbai was the first city in India to initiate a heritage regulation in 1995, with the goal of preserving its historic urban fabric, including buildings, structures, fountains, open spaces, and precincts under various categories.[24] This legislation was partly in response to insensitive repairs and alterations of numerous historic buildings, but most importantly it was a reaction against the early laissez-faire development following the neoliberal reforms of the 1990s. The subsequent creation of the Mumbai Heritage Conservation Committee (MHCC) to implement the legislation created a much needed legal protection for historic sites. (The municipal commissioner, however, retains the power to veto any decisions of the MHCC.)[25] The regulation also created a comprehensive inventory of historic sites in three categories (Grade I, II, and III), with diminishing levels of control around interior and exterior modifications. Today more than six hundred buildings, eighteen precincts, and one open space are part of this heritage listing program. Although the historic designations afford limited protection and are insufficient in managing heritage in dense, low-income, and mixed-use neighborhoods, they are a large step in the right direction toward preserving the city's diverse heritage.

Initially the inventory included several structures in Girangaon as Grade III historic sites. However, the economic frenzy unleashed by the repeal of ULCRA and the relaxation of DCR resulted in a very powerful coalition of capitalist interests, including the builders' lobby, land speculators, and some mill owners. This group collectively wielded far more power than the preservationists, who they caricatured as antiprogress romantics. This coalition saw that the Grade III historic sites were taken out of the purview of the heritage legislation, which has greatly impacted the historic urban environments of Girangaon.[26] Despite a writ petition filed by INTACH in the Bombay High Court, and other citizen groups' public-interest legislations to prevent the loss of mills and *chawls,* their redevelopment continues in various forms.[27]

That piece of the story—that the claims of those with greater economic might often outweigh the claims of the preservation community—is familiar. However, another piece of the puzzle in Mumbai is more unique. A common view of economic globalization, both in India and more generally in Mumbai, sees it as a tool of Western cultural imperialism.[28] Fears of its consequences have triggered a nationalistic wave in much of the country and in Mumbai in particular due to the presence of Shiv Sena. This nationalism has led to an increased interest in the role of history and memory in place making. Ordinarily this would be a boon for preservationists with an interest in retaining the vestiges of history in the built environment. However, both Shiv Sena, and the much larger center-right BJP (Bharatiya Janata Party) have favored a narrow interpretation of history, and the neglect or destruction of heritage when it conflicts with their specific vision of an idealized Hindu past. Some manifestations of this conflict are well known. For instance, in 1992 BJP supporters facilitated the destruction of the Babri Mosque in Ayodhya—a site from the sixteenth century that was contested, as both Muslim and Hindu groups claimed inheritance. In recent years, the parties have successfully renamed historic landmarks with Hindu names (e.g., the British-built Victoria Terminus railway station is now Chhatrapati Shivaji Terminus) and entire cities ("Bombay" to "Mumbai") in an attempt to present a past devoid of any colonial and global histories. Shiv Sena has an additional vision of the hyperlocal; keeping "Mumbai for the Mumbaikars," and the state of "Maharashtra for the Marathis," which means those with long-standing cultural and linguistic ties in the area. The case of Girangaon challenges all of these values. It represents a piece of colonial history, in which working-class immigrants of all religions came from across India (including present-day Pakistan) to Mumbai and toiled alongside the Marathi majority for the British as part of a global economic system. As a result, both Shiv Sena and BJP, despite professing "pro-poor" values, have done little to preserve the heritage of the working classes, which is now fast disappearing in the face of ostentatious developments geared toward the elite. Their approach to "uplifting the poor" hinges on providing

6. Millworkers demand housing and equitable utilization of mill lands in a protest organized by the Millworkers Action Committee. Photograph by Ajit-Bharat/Pukar (Partners for Urban Knowledge Action and Research), Mumbai. Used by Permission.

"free homes" to replace their lost communities. This may have some merit, but its implementation through a neoliberal development vehicle comes at a huge social cost of displacement and alienation. The parties would rather embrace a development model that produces a hypermodern (and symbolically Western) urban aesthetic intrinsically linked to the forces of global capitalistic powers than work on improving layered places that reflect the values of the city's working classes, who are the parties' biggest supporters.

It is important to point out that, even as recent economic reforms and neoliberal urban policies threaten the existence of Girangaon as a working-class industrial neighborhood, at the same time preservation practice in the city as a whole has to some degree benefited from them. With the growing presence of global commerce, there is an ever-increasing availability of funding for restoration projects elsewhere in the city. For instance, various banking institutions (such as Thomas Cook and Deutsche Bank) have spearheaded restoration of buildings in which they are housed as part of their corporate philanthropy mission.[29] Other large Indian corporations such as Tata Sons Ltd. have also funded restoration of historic buildings, including the Bombay House and the Army & Navy Building.[30] However, these efforts have little impact in Girangaon, which does not contain any high-style architectural sites.

The rapid transformation of the city under the dictates of global capital and commerce has also renewed interest in the "local" among some residents who have been directly impacted by the changes. The politics of Shiv Sena is but one extreme form of regionalism; other forms in Mumbai have had more positive consequences for cultural heritage. Among these are increasing activism and citizen participation around redevelopment proposals and preservation projects. Engagement by a broader group of stakeholders and community

7. Scene from the play "Cotton 56 Polyester 84," with a painted backdrop of mills and towers. The play depicts the life of jobless mill workers in Girangaon and the changing fabric of their neighborhoods. Photograph by Sunil Shanbag. Used by permission.

members has, in particular, expanded the realm of preservation practice, as reflected in the preservation of the historically designated Oval Maidan (an open space) and Kala Ghoda (an arts district), which in recent years has become a vibrant cultural node in the city.[31] Another project, Dadabhai Naoroji Road (originally Hornby Road), demonstrates the commitment of local stakeholders in maintaining the heritage streetscape by adopting design guidelines and funding street-improvement initiatives. This particular form of resurgence of the local in the face of the global could have long-term benefits for historic preservation practice in the city.

However, the preservation projects, both those undertaken by large national or multinational corporations and those spearheaded by local community members, are highly concentrated in relatively affluent areas of Mumbai. Thus, despite many obvious benefits, they also contribute to perpetuating the inequalities that arise from preserving only the histories of the rich and powerful, while ignoring the contributions of the working classes. Girangaon retains especially "valuable texts of social and cultural history" and has great potential to inform about the "neglected aspects of urban history and its pluralistic nature."[32] However, it is difficult for those in the preservation community who value Girangaon's contribution to Mumbai's sense of place to fight the powerful forces of development, especially when, on the surface, these forces promise an economic renaissance of the neighborhood that would greatly benefit the working classes. Beneath that surface, the realities of urban transformation continue to favor the elite and marginalize the poor who must either be displaced or remain to deal with an alienating built and social environment that caters to others' needs more than their own.

Memory and Community in Girangaon: An Alternative Vision for a Global Mumbai

There is still time for preservation practice to achieve a greater degree of success in Girangaon. One part of this endeavor requires broadening the approaches used and developing local standards of best practices that are more effective in managing heritage in industrial and low-income neighborhoods. The current preservation discourse in Girangaon has largely centered on sites of old mills without engaging the larger neighborhoods, including the social networks and life and livelihoods they foster. It is important for preservation practice to involve residents to determine how the current neighborhoods meet or do not meet their needs, and to make them aware of the extent to which redevelopment can incur the loss of community that they may currently take for granted. As with all preservation practice, the approach also needs to avoid the other extreme of maintaining the status quo at all costs, but rather be open to upgrading and rebuilding portions of the built fabric as needed.

However, even the best preservation plan for Girangaon is likely to meet strong and united opposition. As Mumbai morphs into a global metropolis, Girangaon is being positioned by a coalition of powerful forces as a district through which they can promote *their* own narrative of globalization — one in which Mumbai looks much like Shanghai, Vancouver, or a host of other cities around the world. These coalitions of builders and politicians stand to gain short-term benefits from the transformation of the neighborhood into a gleaming skyscraper city. To achieve success, preservationists must promote an alternate image of a global metropolis in which history and memory are integral and yield long-term benefits for the city, including its leaders and investors. Other "world-class cities" — both old and new — have demonstrated the social and economic benefits of maintaining historic neighborhoods that foster a shared sense of place and opportunities for residents of diverse socioeconomic backgrounds. Preservation professionals in Mumbai must draw upon the experiences of cities from around the world to develop the case that a truly world-class metropolis cannot be built at the cost of sacrificing its own history or its working-class communities; rather, for a city like Mumbai, these must be seen as strengths that make vital contributions to the achievement of its global aspirations.

Biography

Manish Chalana is assistant professor of urban design and planning at the University of Washington, with affiliate appointments in landscape architecture and the Jackson School of International Studies. Dr. Chalana's work focuses on planning history, historic preservation planning, and international planning and development, particularly in his native India. His current projects focus on sustainability and rural planning in the Himalayas, and the impacts of globalization and urbanization on Indian heritage. He has published in *CRM: The Journal of Heritage Stewardship, Journal of Architectural Education and Preservation Education & Research*. He is currently developing a book manuscript on historic preservation planning in India.

Notes

[1] Jyoti Hosagrahar, *Indigenous Modernities: Negotiating Architecture and Urbanism* (London: Routledge, 2005).
[2] Marc Augé, *Non-Places: Introduction to an Anthropology of Supermodernity* (London: Verso, 2005).
[3] Walter C. Wood, "Effect of the Cotton-Spinning Industry in China upon the Kindred Industry in India," *Journal of the Society of Arts* 46 (1989): 680.
[4] Ibid.
[5] Darryl D'Monte, *Ripping the Fabric: The Decline of Mumbai and Its Mills* (New Delhi: Oxford University Press, 2002).
[6] Neera Adarkar, "Mumbai's Industrial Landscape," in *Mills for Sale,* ed. Darryl D'Monte (Mumbai: Marg Publications, 2006).
[7] Ibid.
[8] Robert K. Home, *Of Planting and Planning: The Making of British Colonial Cities* (London: Spon, 1997), 93.
[9] Collective Research Initiatives Trust, "*Housing Typologies in Mumbai,*" www.crit.org.in/wp-content/uploads/housing_typologies.pdf.
[10] Neera Adarkar and V. K. Phatak, "Recycling Mill Land: Tumultuous Experience of Mumbai," *Economic and Political Weekly* 40 (2005): 5365–68. Rajnarayan Chandavarkar, *The Origins of Industrial Capitalism in India: Business Strategies and the Working Classes in Bombay, 1900–1940* (Cambridge: Cambridge University Press, 1994).
[11] Gurcharan Das, *India Unbound* (New Delhi: Viking, 2000).
[12] Mumbai Metropolitan Region Development Authority, *Summary of the Regional Plan for MMR,* http://www.regionalplan-mmrda.org, vi.
[13] Bombay First and McKinsey & Co, *Vision Mumbai: Transforming Mumbai into a World-Class City: A Summary of Recommendations* (2003): 23.
[14] Darshini Mahadevia, *Inside the Transforming Urban Asia: Processes, Policies, and Public Actions* (New Delhi: Concept Publishing Company, 2008).
[15] Shekhan Krishnan, *The Murder of the Mills: A Case Study of Phoenix Mills—A Report by the Girangaon Bachao Andolan and Lokshahi Hakk Sanghatana* (2004), http://www.heptanesia.net/documents/phoenix.pdf.
[16] Damodar, "Vision Shanghai over the Mumbai Mill Lands—Speculation and Other Ravages of a Decadent Bourgeoisie," *People's March Voice of the Indian Revolution* 7 (2006): 18–24.
[17] David Harvey, *Spaces of Global Capitalism: Towards a Theory of Uneven Geographical Development* (London: Verso, 2006).
[18] Naresh Kamath, "BMC Plans to Revamp British-Era Chawls," *Hindustan Times,* Mumbai Edition, July 10, 2010, http://www.hindustantimes.com/India-news/Mumbai/BMC-plans-to-revamp-British-era-chawls/Article1–568490.aspx.
[19] Rajshri Mehta, "NTC 'Throws Away' 3 Chawls for Rs47 Cr," *Daily News & Analysis,* March 13, 2009, http://www.dnaindia.com/mumbai/report_ntc-throws-away-3-chawls-for-rs47-cr_1238570.
[20] Gwendolyn Wright, "Building Global Modernisms," *Grey Room* 7 (2002): 129. Dirlik, "Architecture of Global Modernity."
[21] Ketaki Ghoge, "Textile Museum Work to Begin Soon," *Hindustan Times,* Mumbai Edition, July 3, 2010, http://www.hindustantimes.com/India-news/Mumbai/Textile-museum-work-to-begin-soon/Article1–566730.aspx.
[22] Manish Chalana. "Slumdogs vs. Millionaires: Balancing Urban Informality and Global Modernity in Mumbai, India," *Journal of Architectural Education* 63 (2010): 25–37.
[23] http://whc.unesco.org/en/statesparties/in. India joined UNESCO in 1946; after almost four decades of no listing, the first two properties were inscribed on the World Heritage List in 1983. Since then, twenty-five additional properties have been added to the list.

[24] Abha Narain Lambah, "Management of Urban Heritage: Bombay, a Case Study," in *Heritage and Development,* ed. AshaRani Mathur, 426–31 (Delhi: Indian National Trust of Art and Cultural Heritage, 2008).
[25] Sharada Dwivedi, "Heritage Committee Is Toothless Tiger," *Daily News and Analysis,* April 14, 2010, http://www.dnaindia.com/mumbai/interview_heritage-committee-is-toothless-tiger_1371123.
[26] Ibid.
[27] Shyam Chainani, *Heritage and Environment: An Indian Diary* (Mumbai: The Urban Design Research Institute, 2007).
[28] Edward Said, *Culture and Imperialism* (New York: Knopf, 1993).
[29] Abha Narain Lambah, "Management of Urban Heritage."
[30] Vikas Dilawari, "Two Decades of Conservation: Lessons from Mumbai," in *Heritage and Development,* ed. AshaRani Mathur, 405–8 (Delhi: Indian National Trust of Art and Cultural Heritage, 2008).
[31] Rahul Mehrotra, "Constructing Historic Significance." *Future Anterior* 1 (2004): 25–32.
[32] Paul Groth, "Generic Buildings and Cultural Landscapes as Sources of Urban History," *Journal of Architectural Education* 41 (1988): 41–44.

1. The residents contributed funds to make this plaque. It was placed at the entrance of Lincoln Lane to introduce to the public the history of the Lane, March 2004. Photograph by the author.

Qin Shao

Citizens versus Experts

Historic Preservation in Globalizing Shanghai

"Mr. Mayor of Springfield, How are you?" reads the opening of an urgent letter dated July 20, 2003, from a Shanghai neighborhood to the mayor of Springfield, Illinois. It appealed for the mayor and the people in President Abraham Lincoln's hometown to help protect "Lincoln's spirit" as well as Sino–U.S. friendship, which, the letter emphasized, were uniquely embedded in their neighborhood but were in imminent danger.[1]

This neighborhood was Lincoln Lane, in the Hongkou district of Shanghai, said to be the only place in this worldly city named after an American president. At risk was not exactly "Lincoln's spirit" or Sino–U.S. relations, but the Lane and the neighborhood, which had been under siege, facing demolition, since fall 2002. The letter was part of a concerted campaign the residents launched to save their neighborhood.

Built in the 1920s on an area of 5,650 square meters, reportedly by a young man who had studied in the United States and who was said to be an admirer of President Lincoln, this group of twenty-five units of three-storied, red-brick row houses, each equipped with modern amenities and a small garden, was by any measure an upscale alleyway residential area that invoked the charm of old Shanghai. A gated community tucked away from the traffic of a main circular—with the historic Hongkou (now Lu Xun) Park established in 1896, a subway station, the Hongkou Football Stadium, a renowned antique market around which modern China's giant literary and political figures once trod,[2] and numerous shops along the famous Sichuan Road, all within its walking distance—it was also a prime piece of property that tempted developers.

In late 2002, the Hongkou district government informed the residents that their Lane was to be demolished for a new subway line to expand the Shanghai metro system. But the residents' own investigations soon revealed that their Lane was to be demolished for the construction of two twenty-eight-story commercial buildings and that they were simply being lied to. Like in many other cases in Shanghai and beyond, the local government used a public project as a pretext to clear out the residents for a commercial development.[3]

The residents immediately started a campaign to protect their neighborhood. They researched the history of the Lane and constructed a useful past to argue that their neighborhood be preserved as "excellent historic architecture," a rare piece

Future Anterior
Volume IX, Number 1
Summer 2012

of material history, for the sake of the city, since quality alleyway houses from the 1920s were fast disappearing in Shanghai. For a moment, they were almost successful.

Ironically, the final collapse of the residents' effort, together with their Lane, came after the government enlisted the support of some renowned preservation experts in Shanghai who apparently saw no value in protecting Lincoln Lane. Among them was Ruan Yisan, a professor who has gained an international reputation as "Preservation God" in China. While he has maintained the rhetoric of an ardent preservationist in public, he and his colleagues' unanimous vote on a government-appointed and paid, closed-door review panel was the last nail that sealed the fate of Lincoln Lane; it fell in late 2004.

What happened in Lincoln Lane was part of a citizen preservation movement that has been on the rise across Chinese cities since the late 1990s. Often scattered, grassroots, and unrecorded, citizen preservation has perhaps been the most intriguing but least studied topic in the field of historic preservation and globalization in urban China. The global economic force behind the breakneck development in China that has destroyed massive tangible historic heritages has also generated new agencies, new dynamics, and new power brokers in the field of historic preservation, which stands at the crossroads of economic development and cultural identity, and of the past and future. The residents of Lincoln Lane, for instance, reached out across the Pacific to Illinois for support of their cause, with an understanding of the global appeal of preservation. Their struggle with the government and dealing with experts such as Ruan Yisan reveals much about the economics and politics of preservation in today's China. Based on field research in Shanghai, this case study of Lincoln Lane sheds light on some of the characteristics of historic preservation in the increasingly globalizing urban China.

The Issue of Preservation

The preservation of historically significant projects in Shanghai started in 1991, on the eve of the rapid development that would remap the city beyond recognition. That year, Mayor Huang Ju issued a directive on "Methods to Preserve and Manage Excellent Modern Construction in Shanghai." It stated that buildings constructed between 1840 and 1949 with the following four characteristics qualified as excellent modern structures: structures were (1) representative of Chinese urban architecture and by famous architects, (2) uniquely artistic in architectural styles, (3) unique in architectural technique, (4) and representative of traditional architectural style and landmarks in Shanghai. The emphasis was on architecture itself. The document placed the responsibility of identifying, preserving, and managing such structures on three relevant government bureaus: the Shanghai Municipal Urban Planning Bureau (SMUPB), the Shanghai Municipal Housing and Land Management Bureau (SMHLMB), and the Shanghai Municipal Cultural Relic Management Bureau (SMCRMB).[4] But the inevitability of the destruction of historic buildings became immediately evident as the large-scale development started in 1992. Two years later, the SMHLMB was compelled to reissue the mayor's 1991 directive and reemphasize the importance of preservation because "damage to preserved buildings has caused broad attention and reaction."[5]

Preservation, however, was a low priority in an era obsessed with growth. Amid public outrage, historic buildings and projects continued to disappear in Shanghai at an alarming rate. In July 2002, the Standing Committee of the Shanghai Municipal People's Congress (SMPC) issued Document 70, titled "Regulations on the Protection of Historic and Cultural Scenic Districts and Excellent Historic Construction in Shanghai," with the most detailed instructions to date on preservation. It indicated that any buildings more than thirty years old with unique architectural styles and representative of the city's history and culture would be considered historically excellent. Most important of all, Document 70 allowed residents to identify and recommend preservation targets to the SMUPB and SMHLMB for research and proposals.[6]

But there were also shortcomings in this and the previous documents on preservation. One problem was that the criteria they provided for preservation were widely open to interpretation and therefore maneuvering by anyone involved; the other is that the documents did not spell out clearly the relationship among the various government agencies involved in the issue, and between those agencies and local governments, as to who had the ultimate authority to make decisions on preservation. These problems, as I will explain, would contribute to the eventual demise of Lincoln Lane in 2004.

But in 2002, Document 70 helped channel residents' voices in preservation. Since 1989 the Shanghai municipal government had launched three batches of preservation projects, which identified a total of 398 buildings and other structures for protection.[7] Municipal agencies such as the SMGLMB, SMUPB, and SMCRMB were involved in the process. In 2002, they focused on selecting candidates for a fourth batch. A team of graduate students from Tongji University, known for its architecture program, was enlisted to help the government agencies in this task by conducting a survey of old buildings in Shanghai generally and by identifying specific candidates for the fourth batch of preservation projects. The student team launched a Historic Preservation Hotline to tap into the city's most enthusiastic preservationists, the residents, since their participation was legitimized by Document 70. The hotline, an eleven-digit mobile number that sounds like, in the Shanghai dialect, "do not demolish [my house]," was a deliberate effort to make a point.

The hotline proved to be a hit. On September 1, 2003, when the local *Xinmin Evening News* reported the hotline, the phone immediately began to ring nonstop and continued well after midnight and overwhelmed the Tongji team. Indeed, for many residents and communities, getting their properties recognized as "excellent historic buildings" was the best and often only way to spare them from demolition. At the time, many of them anxiously hoped that their buildings would be included in the fourth batch of the preservation project. In October, the student team, together with some faculty, held a meeting with those residents to identify projects for preservation. This meeting helped build a community of those interested in preservation and also put the residents in touch with the professionals.

The residents in Lincoln Lane were among those who called the hotline and sent representatives to the meeting. But they came more prepared than most of their fellow citizens. In fact, they had produced a chronicle of the Lane to highlight its significance, which deeply impressed the Tongji student team that eventually came to investigate the history of the Lane.

Citizens in Action

The chronicle was a result of ten months of collective effort by the Lincoln Lane residents. Ever since they were informed of the demolition of the Lane, the residents, anxious and vigilant, had been paying close attention to both municipal policies and district maneuvering. Organized by Mr. D, a charismatic neighbor who emerged as a leader, they saw an opening in Document 70 and decided to recommend their Lane for protection.

They started with research, which proved to be key. What was the history of the Lane? Did anything important and interesting happen in the past? Were there any notable former and current neighbors? The residents, working like researchers, historians, anthropologists, archivists, and writers, documented the history of their Lane as meticulously as possible with oral and printed sources. Mr. D, a retailer with a flexible schedule, made countless trips on his motorcycle to various libraries, bookstores, and archives. When he found a promising lead that needed further research and evidence, he would share the task with his neighbors. Residents with regular jobs went door to door in the evening interviewing older neighbors and family members. Retirees and the unemployed helped make phone calls, sort out interview notes, cut newspaper clippings, and piece the information together. They met every evening to exchange information and figure out the next step. The purpose was to dig out and organize any and all valuable information concerning the history of their lane to strengthen their case for preservation. Gradually, they compiled a chronicle of the Lane with highlights of significant events and people, all annotated with dates, names, eyewitnesses, addresses, phone numbers, articles, and books, and appendixes.

According to that chronicle, Lincoln Lane was nothing ordinary at all. Its survival of the 1932 war was itself a miracle, as its strange numbering system of the units demonstrated—it started with number 11 instead of 1. On January 28, 1932, Japanese troops' bombing of Shanghai killed about ten thousand soldiers and civilians and destroyed many commercial and industrial establishments.[8] The first ten units of Lincoln Lane were among the casualties, but the rest of the Lane was spared only because, the residents, perhaps inspired by its name, flew an American flag on its roof top for protection. In the early 1950s, the Lane was home to the branch office of the Central People's Bank. It suffered under Mao. In the Great Leap Forward of the late 1950s, the two big carved iron doors at the entrance of the Lane became a casualty to the mad rush of neighborhood backyard steelmaking. During the Cultural Revolution (1966–76), the name of Lincoln was seen as pro-Western and redolent of the capitalism that the revolution was intent on wiping out, so the Red Guards changed it to "Miezhi Lane"—literally "destroying capitalism."[9] The original name was restored only after the Cultural Revolution ended.

The Lane even had a direct link to the Chinese Communist Party's own history. Mr. D's home, for instance, was used in the late 1940s by CCP activists, a group of radical students which included D's oldest brother, for underground publishing and networking. That student group was documented in a book titled *Burning Youth* on the CCP's history in Shanghai, a book whose cover title is graced with the calligraphy of Jiang Zemin, onetime Shanghai mayor who was promoted to the chairman of the CCP's Central Committee after the 1989 students' demonstration in Tiananmen Square. The residents in Lincoln Lane placed much hope on this tangible connection between the Lane and the CCP's history because, since the 1950s, the CCP had preserved and restored as relics sites that played a part in its history. Most notable is an alleyway house that hosted the CCP's First National Congress, which had been turned into a sacred museum and is the best preserved alleyway compound in Shanghai.[10]

In addition, the residents presented some illuminating former and current neighbors. They included an American doctor known for reattaching severed limbs; a ninety-two-year-old man who was once the secretary of a former Shanghai mayor;

a teacher of a former premier of China; and the ancestral house of a former finance minister of Hong Kong. The Lane was also said to be linked to Li Guoding, a legendary figure in Taiwan's economic development.[11]

Of these materials, some are confirmed by independent, written sources and others are based on a single, oral account. The vaguest of all was the name of the Lane. Did the name "Linken" in Chinese mean the American president Lincoln, whose name also happened to be translated into the same Chinese characters? Did it reflect the architect's admiration of Lincoln? Who was the architect anyway? These questions remained unanswered. But for the residents, the relevance of their Lane to the American president was self-evident. Indeed, the name was among their most cherished capital, as it allowed them to claim historical and international significance for their Lane.

Based on these materials, the residents made several arguments for preservation. As the Lane had survived China's turbulent history from the 1932 Japanese war to the Cultural Revolution, they argued that it should be at once preserved as a valuable witness to the city's past and celebrated for its strength through enemy attacks and other turmoil. Also, they suggested the Lane be kept as an educational site for future generations to learn about the CCP's history, which is a typical CCP rhetorical move. Furthermore, they pointed out that the Lane, imbued with rich human history, should be protected for Chinese and foreigners who wanted to search for their past. "Standing in front of it, one can feel the history of Shanghai come alive. . . . If the Lane is destroyed, we will regret it for generations to come," they wrote in their recommendation for the preservation of the Lane. They also highlighted the three most important points that made the Lane unique and uniquely valuable for preservation: the Lane's eighty-year history; its well-preserved alleyway style; and its unique name mysteriously linked to an American president.

The residents sent out numerous letters with these arguments to anyone who might help their case, including the authorities, the media, and, as pointed out, the mayor of Springfield, Illinois. In the letter to the mayor, the residents used their imagination to supply the missing links in history and ran with it—they wrote with certainty that the name of their Lane represented "Chinese people's admiration of Mr. Lincoln and support of justice in the world," and that the Lane was thus a "monument that commemorates a world-renowned hero" that must be preserved. What they looked for was some response—any response—from Springfield to validate their claim of a historical link. Such a validation would then strengthen their case and pressure the district—they believed that pressure from a global community would weigh more than pressure from within China—into hearing their case. They tried to send the letter through multiple channels, but no reply ever came.

The residents also took advantage of multimedia for support. Beyond contacting Chinese media and using the Internet, where one of the residents put out a series of "SOS" and appealed to the cyber-crowd to "rescue the old houses,"[12] they also reached out overseas. In fact, the earliest report on Lincoln Lane appeared in an Australian newspaper, *The Chinese Melbourne Daily,* on July 3, 2003, the eve of American Independence Day. The author, an innocent tourist and observer on the surface, gave the essay a deceptively plain but essentially an eye-catching title, "Shanghai Has a Lincoln Lane." It described how impressed he was on his recent visit to Shanghai with the history and architecture of Lincoln Lane. Then he asked a pointed question: "Is this significant and valuable old Lane among those historic buildings being protected in Shanghai?" Not to be too subtle about his own position, he pointed out that since the Lane had survived nearly a century of war and chaos, "It will certainly again show its graceful bearing in the forward looking Shanghai." This was clearly a direct message to pressure the Shanghai officials for preservation, giving weight to the neighborhood's history. This report, it turned out, was initiated in the Lane. A neighbor had a brother working in the Melbourne media and the residents decided to use the opportunity to publicize their case overseas.

To inform the public of the struggle within the rather secluded Lane and garner its support, the residents also raised money to make a special plaque and posted it at the Lane's entrance, right below its address plate. It was titled "A Brief Introduction of Lincoln Lane." This special plaque, eight times larger than the regular, official address plates, with President Lincoln's well-known profile in the background, states the key information about the Lane that the residents wanted to convey: that the twenty-five units of the Lane were built in the 1920s; that the Lane is the only place in Shanghai named after an American President; that CCP underground activities took place there; that there was a Jiang Zemin connection; that its name was changed in the Cultural Revolution and restored thereafter. The residents hoped to use this plaque to publicly legitimize their claim of the Lane's value for preservation. Usually, the municipal government would issue a plaque as a certificate to structures officially identified as "excellent historic buildings." But now the residents took the initiative of putting up their own plaque in the public space to both compete with and challenge the authorities over the status of the Lane.

The residents' campaign attracted the Tongji student team to get involved in the case. Upon careful investigation, it concluded that the Lane belonged to the alleyway type and was a very well preserved garden residency, rare in Shanghai, and worth protection. Their finding brought about more media attention and several news outlets reported on the case. In December 2003, at a SMUPB meeting to compile a provisional list of candidates for the fourth batch of preservation projects, the case of Lincoln Lane was favorably reviewed. In February 2004, the SMUPB and SMHLMB issued Document 107 that recommended the Lane to be included in the fourth batch of preservation projects, pending final municipal approval. The document also indicated that the local government, the Hongkou district in this case, should take active steps to protect it first, as the procedure to finalize its status was taking place.

The Role of the Experts

But the Hongkou district government kept Document 107 a secret from the residents. In fact, while municipal agencies such as SMUPB and SMHLMB were considering the preservation status for Lincoln Lane, the district had worked overtime to destroy the Lane to push forward the construction of the commercial twin towers. Residents suspected that the district had already made a deal with the developer to sell their neighborhood, which has been a common practice in China—the joined force of local government and developers, of power and money, has been responsible for driving real estate development in China. In any event, throughout 2003, the district's demolition squad constantly harassed the residents, forced a dozen families out, tore down their buildings, and twice initiated violent attacks on the residents when they refused to move out. To the residents, the district meant to do enough damage to render their preservation campaign meaningless.

Document 107 thus presented a challenge for the Hongkou district government: how to deal with the municipal agencies on the one hand and the residents on the other. It decided to simply ignore both by stalling the former and keeping the latter in the dark. When the residents finally found out about Document 107 and contacted the media to pressure the SMUPB, SMHLMB, and the district government to reinforce the document, the officials realized that they were in trouble.

But in the one-party, opaque government of China, official conflict is always kept internal, in order to maintain its unity in public. In this case, the various municipal agencies and the Hongkou district must appear to be consistent over the Lincoln Lane case, especially now that they were under public scrutiny. The problem, as mentioned earlier, was that the relationships among those offices when it came to decision making were murky. The district government, with executive authority, was not subordinate to the other municipal agencies. Moreover,

2. The beginning of demolition in Lincoln Lane. January 2004. Photograph by the author.

Lincoln Lane was under the direct control of the district, not the other agencies. As a result, the SMUPB and SMHLMB yielded to the district and had to backtrack on Document 107, to officially disqualify Lincoln Lane as a valuable historic structure and thereby justify its demolition. For that purpose they got the experts involved.

On June 30, 2004, the SMHLMB convened an expert panel to review the Lincoln case and determine, on the surface, whether to recommend it for preservation—but everyone on the panel understood it was to kill the case. Of the twelve persons that consisted of the "Experts Review Committee," six were professors from Tongji University, known experts on architecture and preservation, and the other six associated with the SMHLMB and SMUPB.

Of the panelists, the best known was Ruan Yisan, the seventy-year-old professor and director of the Research Center for Nationally Famed Historic and Cultural Cities at Tongji University, the man of the hour on all things about preservation in China. He was widely credited with transforming some historic towns into commercially successful tourist destinations in the Yangzi region. Leading architects such as Rem Koolhaas have criticized the nature of this type of preservation, which contributes to "gentrification and social displacement" by "driving out the poor to make room for wealthy homeowners and tourists," and which distorts historic structures by "creating tasteful scenery for docile consumers."[13] Regardless, considered to be a leading champion of preservation, or even "Preservation God" in China, Ruan in 2003 was given the UNESCO Asian and Pacific Area Cultural Heritage Protection Outstanding Achievement Award.

The evaluation form for the Lincoln Lane case listed seven criteria by which the panel was asked to render their opinion. They included: whether the structures were unique in architectural style, whether Lincoln Lane reflected the historic and cultural character of the Shanghai region, whether any of its buildings were by renowned architects; how well the site fit into its surroundings, as well as an overall review of the structure. The experts' answers to these criteria were extremely brief, with only a couple of sentences devoted to each, and most criteria were addressed with just a simple "yes" or "no." Of the twelve panelists, only three indicated that Lincoln Lane had a certain uniqueness in its architectural style. Seven did not know anything about Lincoln Lane at all. In the end, only one suggested that the Lane be partially preserved and its "historical memory" be sustained by putting out a plaque.

There were several problems with this so-called expert evaluation. First, the panelists were not required to support their view with any evidence. The explicit reason that they were consulted was their supposed knowledge about such subjects. But most of them knew nothing about the history of Lincoln Lane and its architectural style, and they simply said "no," "not known," or "no details." One panelist even stated that Lincoln Lane was built in 1937, completely disregarding the well-documented fact that the Lane suffered the Japanese bombing in 1932. Willful ignorance and stunning obliviousness were apparently accepted as a virtue or, in a strange way, as a sign of authority. The carelessness with which the experts handled the Lincoln Lane case was in stark contrast to the thoroughness of its residents. The latter, conducting themselves like trained experts, researched their case and documented their argument with oral and printed material and with appendixes. But the actual, trained experts were not required to do the same.

Among the panelists' votes, Ruan Yisan's was the most troublesome. He had publicly expressed his dismay at the wholesale destruction of old Shanghai. At the October 2003 meeting at Tongji University, where he met residents from Lincoln Lane and elsewhere, he stated that there were two thousand alleyways in Shanghai prior to 1949 but now fewer than one thousand left, and every one of them should be preserved. Then in November, regarding specifically the Lincoln Lane case, Ruan was still adamantly against demolition. One point he made in one of his interviews was perhaps the most relevant: "If experts, residents, and the government work together, we can preserve some fast-disappearing examples of our cultural heritage. Our job is to serve the people and help them in recommending excellent historic buildings."[14] Ruan's statements were a godsend to the struggling residents who treated him as a powerful ally and used his remarks to strengthen their argument for preservation.

Yet at the committee meeting, Ruan dismissed Lincoln Lane's significance and failed to recommend it for protection. He considered the Lane's buildings to be "ordinary . . . without historical value or special architectural character," which directly contradicted his own public view of only a few months earlier.

But Ruan's position is perhaps not surprising, considering the fact that he had also capitalized on his fame by establishing a contract firm to work on commercial preservation projects, which largely depended on the government's favor. In other words, he needed local officials' support for his firm. During the Lincoln Lane controversy, he was actually working on a project to preserve the wartime Jewish ghetto in Hongkou, which was granted to him by the district government.[15] His self-appointed, publicly announced responsibility to help people in the cause of preservation suddenly no longer mattered. In the end, "Preservation God" Ruan Yisan was himself a developer after all, driven by the same impulse to pursue profit.

To be sure, Ruan's government connection and hypocrisy were not unique. All the committee members, as known preservationists, were appointed and invited to the panel by the SMHLMB that also paid for their service, 1,000 yuan in an envelope for their brief answers to the seven questions, perhaps the easiest 1,000 yuan anyone could make. The officials' trouble with the Lane and their changed position from Document 107 were not a secret to the experts. If those experts wanted to be invited back for future consultation and continue to profit from their expertise, they had to be in good standing with the officials. While they spoke out openly on preservation, they looked the other way when their own interests were at stake. They had essentially sold out the residents and helped the government destroy their valuable houses to advance their own interests. In the process they also played a hand in wiping out part of the city's irreplaceable history—the large-scale destruction in Shanghai could not have happened without those experts' collaboration with the government.

The panel's review justifiably outraged the residents. In a lawsuit the residents filed afterward against the SMUPB for failing to enforce Document 107, they contested the panel's conclusion and bitterly complained that the experts "knew nothing about history and even showed no interest in it." They called on Ruan Yisan and questioned the fairness of the panel: "If some of them were under contract to the Hongkou district government, they should have excused themselves from the panel." The residents were convinced that "those experts had betrayed themselves and their ruling was unconscionable." They ridiculed one panelist's mistake that Lincoln Lane was built in 1937. The same person also argued that the Lane had no relationship to Lincoln whatsoever, to which the residents responded: "If the Lane was named 'Lincoln,' then how is it not a reference to the American president? Or did the expert have any other explanation?" The residents also took their criticism of especially Ruan Yisan into cyberspace, since he was labeled as a "public intellectual" in the blogosphere. One blogger suggested stripping Ruan of his title as a "public intellectual" because "He is a politician who colluded with the Hongkou district government and sold out Lincoln Lane."[16]

As expected, the panel reviews became the last nail in the coffin for Lincoln Lane. They provided cover for the SMUPB and SMHLMB to retract Document 107 and disqualify the Lane for preservation, and they legitimated the district government's action to evict the residents and demolish the Lane, a process that was complete by the end of 2004. During that time, the subway line was completed without going through Lincoln Lane, and late 2005 witnessed the laying of the foundation for the twenty-eight-story twin towers, which were on the market for sale in 2007. Both the subway line and the twin towers attest to the corruption in Shanghai and the challenge faced by historic preservation. In 2008, however, a single one-story house was built next to the twenty-eight-story high rises as the district government's effort to symbolically "preserve" Lincoln Lane. This new structure only begs the question of why the Lane was leveled in the first place, if it warranted such an awkward remembrance after all.

As for the residents, some of them made a deal with the district and settled elsewhere; others, especially those who were violently evicted, have continued their petitions and lawsuits in an effort to reclaim their homes. The end of the case is not yet in sight.

Concluding Remarks

Historic preservation, like environmental protection, is by and large an issue of the postindustrial age—when the damage is already done, when economic growth is no longer the only concern, and when society has enough resources to invest in preservation. However, despite the vastly different developmental stages among nations, historic preservation, again like environmental protection, has become a global issue. The rapidly shrinking world and the relentless march toward the future, often at the expense of the past, have heightened awareness of mankind's shared heritages across the globe and in the particular cultural heritage and identity based on nation, people, community, and family. But how to deal with this issue continues to vary depending on the developmental stage of the nation. For instance, by spring 2004, Shanghai had designated 398 buildings for preservation. Yet Singapore, the city-state that is the smallest country in Southeast Asia but much more developed than China, has put more than 5,300 buildings under protection.[17]

It is therefore hardly surprising that in the post-Mao, industrializing China, preservation is not a high priority. Rampant corruption, loopholes in regulations, and weakness in the enforcement of the law, which are some of the trappings of an authoritative government in a developing country, have only compounded the problem. More broadly speaking, since its inception, capitalism has marched forward at the expense of numerous historic heritage sites worldwide. What has happened in China is the norm instead of the exception. The politics of preservation, after all, boils down to money. In 2002 when the demolition started, the residents in Lincoln Lane were offered 4,500 yuan per square meter to relocate. In early 2007, the twenty-eight-story building was marketed at 30,000 yuan per square meter. This price gap is what ultimately motivated the district and the developer to destroy the Lane. Market forces rule. Lincoln Lane's fate is shared by many much grander historic landmarks across urban and rural China.

The influence of the market corrupts Chinese intellectuals as well. In fact, "public intellectual" is a contradiction in terms in the Chinese context. The imperial system in traditional China produced a scholarly class whose highest calling and sole mission in life was precisely to serve the emperor. Dissenting from the political center was perilous for the literati, and such was the case throughout Chinese history, including during the Mao era. The post-Mao reform has created some space for independent-minded intellectuals to grow, but they remain a rarity, as their own interest is often tied with that of the government. Ruan Yisan is right in his statement that historic preservation could benefit from the combined effort of experts, government, and citizens. He capitalized on the global trend of historic preservation and built his reputation as an internationally renowned preservationist. But that global profile has not compelled him to act on his publicly stated belief when it was inconvenient for him. As a for-profit developer himself, Ruan, too, yielded to the power of money and interest at the expense of preservation.

What is worth emphasizing in the case of Lincoln Lane is how the local and global dynamics at play at the grassroots level led to the citizens' initiative and participation in historic preservation. To be sure, the residents had their own pragmatic concerns as well. They initially hoped to preserve their homes through the preservation of the entire Lane. But at one point, they also expressed the willingness to relocate if the district could protect the Lane for its independent value. While they were attached to the Lane as a physical embodiment of their family history and memory, they also recognized its importance as part of the city's history. Moreover, as Shanghai has become increasingly an international city and its residents among the most well-informed Chinese citizens, the residents in Lincoln Lane understood the global significance of preservation generally and of Lincoln Lane particularly. They constructed a useful past for the Lane and tried to rally support from America to Australia, as well as from the Internet and other media outlets. Clearly, globalization has provided the residents with not only a heightened awareness of but also more resources to argue for historic preservation. But it has its limit. The Shanghai government took an important step to include the citizens in its preservation effort, but that effort is unlikely to produce any tangible results if the pursuit of economic growth continues to crush everything else in its way.

In the end, the Japanese invasion damaged Lincoln Lane, and the Cultural Revolution renamed it, but it was the triumph of a capitalist economy in the post-Mao era that destroyed it forever.

Biography
Qin Shao is professor of history at The College of New Jersey. She is the author of *Culturing Modernity: the Nantong Model, 1890–1930* (2004) and has published in the *Journal of Asian Studies* and *China Quarterly*. She is currently completing a book project, tentatively titled *Shanghai Gone: The Destruction of Everyday in China, 1980–2010*.

Notes

This article is part of a book project on housing reform and conflict in post-Mao China. A Research Fellowship from the American Council of Learned Societies in 2003 funded the initial fieldwork for this project. In 2007–8, a residential fellowship from the Radcliffe Institute for Advanced Study at Harvard University, a Scholar Grant from the Chiang Ching-Kuo Foundation, and minigrants and other awards from The College of New Jersey allowed me to write and revise the project. I am extremely grateful for all of their support. I also wish to thank Ijlal Muzaffar for his very helpful comments on a draft of this paper.

[1] Unless otherwise indicated, all the unpublished material in this study is based on my own interviews, observations, and investigations directly onsite and with the residents. I conducted my fieldwork in Shanghai in spring 2004, summer 2005, 2007, and 2009, and have also carried an ongoing investigation through e-mail and telephone.

[2] The Duolun Cultural Street is deemed as a must-see for foreign tourists; see http://www.transindus.co.uk/index.php?pg=1&rn=China&t=CCD&td=2.

[3] See another such example in Qin Shao, "Waving the Red Flag: Cultural Memory and Grass-Roots Protest in Housing Disputes in China," *Modern Chinese Literature and Culture* 22, no. 1 (2010): 197–232.

[4] SMHLMB, ed., *Shanghai youxiu lishi jianzhu baohu falu fagui wenjian huibian* (A collection of law and documents on preservation of excellent historic structures in Shanghai) (Shanghai: SMHLMB, 2005; hereafter *A Collection of Law and Documents on Preservation*): 82–84.

[5] http://sh.focus.cn/news/2003-03-13/42523.html.

[6] Shanghai Municipal Housing and Land Management Bureau, ed., *A Collection of Law and Documents on Preservation*, 4–14.

[7] *Xinmin Evening News,* March 18, 2004.
[8] Donald A. Jordan, *China's Trial by Fire: The Shanghai War of 1932* (Anne Arbor: University of Michigan Press, 2001): 188–98.
[9] For more information on the Great Leap Forward and the Cultural Revolution, see Jonathan Spence, *The Search for Modern China* (New York: Norton,1999): 544–53, 565–86.
[10] Hanchao Lu, "'The Seventy-two Tenants': Residence and Commerce in Shanghai's *Shikumen* Houses, 1872–1951," in *Inventing Nanjing Road: Commercial Culture in Shanghai, 1900–1945,* ed. Sherman Cochran, 174–77 (Ithaca, N.Y.: Cornell University Press, 1999).
[11] See more on Li Guoding at http://blog.sina.com.cn/s/blog 4b8bd145010082af.html.
[12] http://hrb.focus.cn/msgview/4059/50850005.html; http://www.abbs.com.cn/bbs/actions/archive/post/2701329 0.html.
[13] http://www.nytimes.com/2011/05/24/arts/design/cronocaos-by-rem-koolhaas-at-the-new-museum.html?emc=eta1.
[14] *Youth Daily,* November 5, 2003.
[15] In 2005, the wartime Jewish residential area of seventy acres was designated for preservation. But in early 2009, part of the buildings in that area were threatened with demolition for a road project. Ruan Yisan who was involved in the preservation project, was protesting the decision, perhaps this time for real. See http://www.npr.org/templates/story/story.php?storyId=100522156.
[16] http://vip.bokee.com/authorEdit/articleCommentAll.php?id=44050.
[17] *Wehhui bao,* February 5, 2004.

1. Interior of Gurdwara Shahid Burj Bhai Jivan Singh, Chamaur, Punjab, India, 2008. A panel of photographs depicting the demolition of the original building on this site is prominently displayed on an interior wall. The middle photograph of the panel depicts an old well in the foreground, which was incorporated into the ground floor of the new building. Photograph by author.

William J. Glover

Shiny New Buildings

Rebuilding Historic Sikh Gurdwaras in Indian Punjab

Sikh religious architecture may have first received notice as a sui generis object of architectural historical interest in amateur British historian James Fergusson's 1892 book, *History of Indian and Eastern Architecture.* Despite the book's claim to being "quasi-exhaustive," here is almost all that Fergusson wrote about Sikh architecture:

> It does not . . . seem that temples or gorgeous ceremonial formed any part of the religious system propounded by [Sikhism's] founders. Reading the 'Granth' [holy book] and prayer are what were insisted upon, but even then not necessarily in public. We, in consequence, know nothing of their temples, if they have any.[1]

Fergusson then went on to describe the Golden Temple in Amritsar (which he presumably knew at least something about) in three sentences:

> Runjeet Singh was too emulous of the wealth of his Hindu and Moslem subjects in this respect not to desire to rival their magnificence, and consequently we have the Golden Temple in the Holy Tank at Amritsar—as splendid an example of its class as is found in India, though neither its outline nor its details can be commended. It is useful, however, as exemplifying one of the forms which Indian temple-architecture assumed in the 19th century, and where for the present, we must leave it . . . now that the sovereignty has passed from the Sikhs we cannot expect their priests or people to indulge in a magnificence their religion does not countenance or encourage.[2]

Fergusson's prediction that Sikhs would probably not produce an architecture of magnificence due to strictures inherent in their faith seems misguided in light of the more or less wholesale refashioning of important historical gurdwaras (and the prolific construction of new ones) that has taken place over the last hundred years or so in India, a process that has produced a considerable amount of "magnificence" along the way.[3] While this robust building activity points to architecture's renewed importance as a focus of Sikh community attention (and, in particular, of philanthropy), the manner in

Future Anterior
Volume IX, Number 1
Summer 2012

which historic gurdwaras have been rebuilt in South Asia is the subject of considerable debate, both within and outside of Sikh communities. This debate often pits enthusiastic lay volunteers, building contractors, and donors, who rebuild historic sites using up-to-date construction materials and methods, against academically trained preservationists, historians, and heritage activists who argue for the preservation of older sites intact and who sometimes deplore the work of the former groups as crass, destructive, and misguided.

At first glance these different attitudes appear to map comfortably onto a well-established conceptual opposition between those who privilege "heritage" as the appropriate mode for apprehending the significance of the past versus those who privilege "history."[4] The first term refers to the past as a quality immanent in the present, something that may take the form of a legacy or inheritance that endures over time. Heritage, therefore, presupposes continuity through time and thoroughgoing connections between past and present experiences. "History," conversely, refers to the past as that which happened prior to the present and whose complete recovery is no longer possible. For this reason, history only ever "recovers" the past as a series of discrete fragments, each of which is subject to objective verification. The historian's work therefore is largely one of interpretation: establishing connections between objectively verified fragments, among which physical artifacts have long played an important role.[5] Both those who privilege "heritage" and those who privilege "history" make claims to represent the past "as it really was," but each establishes the veracity of her claims through recourse to different modes of argumentation. For the historian, veracity is established on the basis of what she believes to be sound, objective evidence. For heritage enthusiasts, according to anthropologist Valentine Daniel, veracity lies "not in any particular object or objectivity, but in a lived experience of intersubjective being."[6]

Applying these terms to our context we may say that those who enthusiastically demolish old buildings on important sites in order to replace them with newer, purportedly better ones are able to conceptualize the destruction they cause as incidental, since the object of their labor is to strengthen the quality of Sikh "heritage" as something vibrantly alive in the present. For those who decry the destruction of older buildings, however, each demolition incrementally diminishes the possibility of knowing what the past—as "history"—actually consisted of. Yet the statements and activities of actual groups are often difficult to reduce to the simple distinctions between "heritage" and "history" outlined thus far. Instead, gurdwara demolition and rebuilding in India today encompasses a range of complicated processes that defy such categorizations. As

I hope to suggest, peoples' attitudes toward the Sikh past hold both history and heritage in sharp focus, and the material landscapes that embody and annotate those attitudes make space for both positions simultaneously, though seldom without controversy.

Shiny New Buildings on Old Sites

Gurdwara reconstruction is passionately engaged in by Sikh *kar sevaks* and the organizers, patrons, and officials who help pay for and administer new construction. While the practices and attitudes of groups involved in gurdwara reconstruction are diverse, and no single group can speak for the majority, the practice of *kar seva,* which ties them all together, has a long history in Sikhism. Originally intended to vest the responsibility for maintenance, upkeep, and construction of property in the hands of the *sangat* [congregation] rather than a particular individual (who may be seen thereby to improperly benefit from the work), *kar seva* activity has become increasingly politicized, both reflecting and advancing differing ideas of what constitutes the Sikh *panth* [community]. Still, to participate as a *kar sevak* is both a duty and a privilege for most Sikhs, and the demolition of older gurdwaras in order to replace them with new, larger, more finely finished buildings brings considerable satisfaction to a great number of people.

Evidence for this is not difficult to confirm. At Gurdwara Shahid Burj Bhai Jivan Singh, for example, a building located on a promontory in the town of Chamkaur Sahib (a site that was built to commemorate Bhai Jaitha, who carried Guru Tegh Bahadur's severed head back from Delhi to present it to the latter's son), photographs documenting the destruction of the original building by *kar sevaks* were prominently and proudly displayed on a series of panels in the hall of the new building that replaced the old one. The smiling men, women, and children engaged in demolishing what appears to be an eighteenth- or nineteenth-century structure present visual testimony of their conviction about the pious and useful nature of the work. The caretaker of the new building, which was still under construction when I visited it in June 2008, showed me the new building with an obvious sense of pride. When I asked him why a new structure was built rather than simply repairing and preserving the old one, he answered that the old building was made of inferior material, was unsafe, and was far too small.

Indeed, the relatively humble nature of historic gurdwaras is often cited as a root cause of their precarious condition. Delhi-based conservation architect Gurmeet Rai and art historian Kavita Singh argue that the relative fragility of older Sikh buildings is an important reason why so few remain standing today. "For the most part," they write, "the building projects of the Gurus were modest ones, made through the collective efforts and donations of their followers. As a result, historic structures from the period of the Gurus were unlikely to survive the vicissitudes of time."[7] At the same time, Rai and others have blamed inadequately trained and supervised *kar sevaks* for destroying what time alone could not. According to Rai, as many as "[ninety percent] of Sikh heritage monuments have been destroyed in Punjab in the name of renovation and *kar seva*."[8] Rai sees the problem as stemming from peoples' ignorance about the historical significance of physical remnants of the past: "ironically," she writes, "most physical evidence of the past has been either bulldozed out of existence or changed beyond recognition by well-intended, though ill-informed keepers of these historic sites."[9] In Rai's view, rebuilding older sites rather than preserving them intact is a form of what she calls "vandalism from within [the community]." "The past has to be conserved not reinvented," Rai writes, "it cannot be deconstructed to reflect today's insensitive attitudes. Every old remnant of the past has to be preserved to reflect the mood, materials, construction techniques, forms and functions of those times. To rebuild [historic gurdwaras] in inappropriate and bizarre materials like marble, or to obliterate them altogether, is to slam the door in the face of our history and heritage."[10]

The Sanctity of Sites versus the Sanctity of Monuments

To understand more fully what is at stake in these debates we have to look more closely at what actually takes place in the renovation projects themselves. While in most cases it is true that much of the physical fabric of the older buildings is demolished, there is usually some portion of the original site retained in the new structure as well. Consider the following example. In his book *Sikh Architecture in Punjab* (1986), architectural historian Pardeep Singh Arshi published an old (undated) photograph of the Guru ka Lahore gurdwara near Anandpur, a small single-celled building with distinctive pilasters at each corner and repeating blind arches on each façade.[11] During the late 1980s, when Arshi was completing fieldwork for his doctoral dissertation at Guru Nanak Dev University, this building was being rebuilt. A new two-story building on the site is considerably larger than the original one, with an open gallery on the second level. The large masonry structure that covers the *palki* [canopy covering the Guru Granth Sahib, or holy text] inside the new building may in fact comprise the remains of the original building. Though the photograph in Arshi's book is too poor to be able to confirm this with certainty, the inner chamber is roughly the same size and scale as the original building and

2. Interior of Gurdwara Guru ka Lahore, near Anandpur Sahib, Punjab, India, 2007. The masonry structure in the center of the photograph may be partially constituted by an earlier structure on the same site. Photograph by author.

reproduces the older building's corner pilasters. If this hypothesis is speculative, however, then a number of other rebuilding projects provide conclusive evidence that even when most of the original structure is demolished to make room for a new building, some core feature of the original structure (and perhaps the entire original structure) will often be preserved in the new one.

At Gurdwara Shahid Burj Bhai Jivan Singh in Chamkaur, a building mentioned briefly above, an older well attached to the original building is preserved intact and given a prominent place in the ground floor of the new building. A similar feature can be found at Gurdwara Shri Patshahi Nauvin Qila Bahadurgarh just outside of Patiala. Demarcated by tradition as the site where Guru Tegh Bahadur came to meditate daily while he stayed with Saif Ali, an old well and the original *takht* [platform] that the Guru was said to sit on during his meditations have been incorporated into the ground floor of the new structure. Examples like this could be multiplied, but these few will suffice to suggest that even in notoriously destructive *kar seva* projects, there exists some component of reverence for "the mood, materials, construction techniques, forms and functions of" older buildings in the sense that Gurmeet Rai finds missing.[12] In each of these examples, however, only the core of the original building is preserved. The rest of the building simply protects the site and facilitates worship by the *sangat* there according to their liturgical priorities.

3. Gurdwara Shri Patshahi Nauvin Qila Bahadurgarh inside the precinct of Bahadurgarh Fort, near Patiala, Punjab, India, during construction in 2007. The large round structure in the foreground is a newly constructed appendage to an old well. In the background, beneath a brick canopy with cuspate arches, is a brick platform said to be the site where Guru Tegh Bahadur sat and meditated. Photograph by author.

This practice raises the question of why some portions of the older building sites are preserved in reconstruction activity, while other portions are not. Historian Gurinder Singh Mann has argued that "the sacrifice of life for the protection of the Sikh community translated into the emergence of spots soaked in Sikh blood as part of a shared sacred geography. . . . The touch of the Guru *(than suhava)* and the shedding of Sikh blood *(shahid ganj)* came to mark the sanctity of a spot, and this sense has persevered."[13] Most of the important historical gurdwaras in India appear to function in this way: to mark the place where something important happened in the history of the Sikh community, particularly in relation to the Gurus and their families. We may loosely think of these buildings as martyria, therefore, erected in particular, irreducibly specific sites, as opposed to those gurdwaras erected by *sangats* for purposes of communal worship on sites chosen for reasons of economy, proximity, and so forth, and that do not mark the spot at which something or someone holy literally "took (their) place." The reverence for material traces of these holy spots for many, or perhaps most Sikhs, does not extend to the entire building fabric of the original (or later) gurdwaras erected to commemorate them. And yet this is not quite the end of the story, since the new buildings going up over old sites are themselves entangled with history in complicated ways.

Critics of newly built structures inevitably single out their use of glitzy materials, including marble, mirrors, and decorative ceramic tiles, as being inappropriate to their settings. Yet almost none of the rebuilt historical gurdwaras have adopted anything but a "period" architectural style for their overall composition and decorative schema. Columns, domes, *chhajjas* [projecting eaves], and *chattris* [umbrella-like domelets] of the kind found on sixteenth- and seventeenth-century religious buildings or elite *havelis* all across north India are turned out at these sites by the dozens with the aid of prefabricated molds and templates. In addition to glossy ceramic tiles and white marble cladding, a number of newer buildings have taken ornamental cues directly from Harmander Sahib, or the Golden Temple in Amritsar, the early nineteenth-century building described by Fergusson, above. However much critics of these new buildings may decry the damage they inflict on older structures, their present-day builders cannot be accused of ignoring the opportunities presented by the rebuilding process for expressing a historical sensibility, one that does not so much "slam the door on" the Sikh past as proliferate the modes through which it is tangibly expressed.

The modes are multiple. One of the most conspicuous features of recent alterations to historical Sikh sites is the almost excessive amount of written annotation that accompanies them. Exhaustive lists of donors have long been prominent at these sites, often found as inscriptions on the surfaces of buildings and pavements. More recently, placards, posters, publications, and didactic exhortations of all sorts have become ubiquitous at every major site. Important historical spots are annotated by placards (usually written in both Punjabi and English, a practice that underscores the importance of foreign or expatriate visitors to these sites) explaining what happened to whom, where it happened, and when. Increasingly, reproductions of paintings about the events being commemorated by the architecture are placed in direct contact with buildings as well, providing a didactic supplement to the more subtle "period" atmosphere posited by the building's architectural features. Put simply, these are material landscapes that work in multiple ways, and with diverse media, to continually remind visitors of the way the past lives on in the present, and which are rarely satisfied to locate "pastness" as a quality inherent only in older objects or people. Note that this doesn't necessarily diminish the reverence for past persons and objects, either: this is how we should see the preserved wells and *takhts* I referred to above, or indeed the carefully preserved old coins, arms, and personal items that are often collected and displayed at these sites. How, then, should we explain this

4. Interior of Gurdwara Qatalgarh Sahib in Chamkaur, Punjab, India, 2008. Two backlit electric display boxes illuminate large-scale color transparencies of paintings that depict the battle of Chamkaur in 1704. This building commemorates the battle. Photograph by author.

bifurcated attitude toward the past, one that simultaneously valorizes both architectural "heritage" and objectively verifiable "history"?

Professor Anne Murphy has studied eighteenth-century *gurbilas* literature composed just at the time when the substance of the last living Guru was transferred to the text of the Guru Granth Sahib and to the Sikh community at large. Murphy argues that these developments helped produce a distinctive sense of historicity within the Sikh community that entailed a kind of doubling—the past adhering to things and people that are either very old or long departed, on one hand, and the past as an "everpresent" quality whose locus is the community itself, on the other. The transfer of Guruship from the living man to the nonhuman text, Murphy argues, hinged importantly on the idea that "the presence of the Guru continues beyond his death right up to the present and beyond. This initiates a new relationship of the community with its past . . . a new temporality, and therefore a new historical sense, one in which the past is in one sense gone, and in another [sense] everpresent."[14]

Twentieth-century rebuilding projects don't simply reflect a Sikh historical sensibility that hasn't changed since the eighteenth century; rather, the evolution of the Sikh panth enabled a distinctive historical sensibility to emerge over time whose concrete expression in any given period depends on what resources are available to it. In the context of gurdwara rebuilding in the twentieth century, we would have to include among those resources an increasingly diverse and wealthy Sikh diaspora, some of whose members direct their philanthropy towards building projects in India (though not only religious buildings); the early twentieth-century vesting of gurdwara management in the institutional form of the Siromani Gurdwara Parbandhak Committee (SGPC) and thus the production of an "official" past that these buildings must respond to and annotate; and, more recently, the proliferation of what we might call postliberalization materialities accompanied by an aesthetic that valorizes marble cladding, glossy finishes, and exorbitant embellishment of all sorts for use in a range of middle- and lower-class residential, retail, and religious structures, regardless of ethnic or religious identification.

We also have to take into account the growing spread and popularity in the twentieth century of museums, in their multiple forms, as privileged sites for the presentation and ordering of history in India.[15] Many gurdwaras, if not most, format historical materials on site using principles and devices adopted directly from museum exhibition practice. At Gurdwara Qatalgarh Sahib in Chamkaur, for example, a wood-and-glass case displays historical weapons and musket shot uncovered in excavations during the building's reconstruction. The blue velvet lining, artful arrangement of objects (each isolated from the others) and descriptive annotation on the exterior of the case all derive from standard museum designs in use in India since the late nineteenth century.

History, Heritage, and Museums in Anandpur

In the town of Anandpur, India, this last line of inquiry bears particular relevance. Renowned as a site where the Sikh Khalsa [community of Sikhs who have undertaken formal initiation rites] was inaugurated by the tenth and final living Guru, Gobind Singh, Anandpur is replete with important commemorative sites and structures.[16] During the nineteenth century, Anandpur's five original forts, built by Guru Gobind Singh in the late seventeenth century, were commemorated by building gurdwaras at each. Other important gurdwaras in the town were built on sites commemorating important events in the lives of Guru Tegh Bahadur, his son Guru Gobind Singh, and their families, including the site in Guru de Mahal where Guru Gobind Singh was installed as the tenth guru (Damdama Sahib), Gurdwara Sisganj; a building that marks the place where Guru Tegh Bahadur's severed head was cremated by his young son (in 1675); the site where the latter addressed Sikhs gathered for the cremation of his father (Akal Bunga); and the site where a number of Sikhs were killed during a siege by Mughal generals in 1705 (Shahidi Bagh); among others. However, there are at least two important buildings in Anandpur that have little organic connection to the sacred geography of Sikhism but whose presence nevertheless works toward supporting the diverse meanings that geography offers to Sikh communities by other means.

The first is the Shri Guru Teg Bahadur Museum, designed by architect Surjit Singh in 1979 and completed in 1983. This building is located across the road from Sri Keshgarh Sahib (one of the five *takhts* or seats of Sikh panthic authority) on a site overlooking a large field, parking lot, and water tank lying to the west of the built-up part of town. Unlike the rebuilt gurdwaras in Anandpur that commemorate the life and activities of the gurus, many of which were rebuilt around the same time that the museum was constructed, the Guru Tegh Bahadur Museum is designed in a style thoroughly in keeping with the international architectural trends of its time. The building presents a mostly blank face to the city, and its light brick walls emphasize sculptural mass and solidity rather than an identifiable Indic or historical style. A pleated second story rises from the ground floor block and terminates in a projecting concrete cornice. A large translucent dome in the center of the roof illuminates a staircase below. The only other openings in the building's exterior walls are the front entry and strips of narrow clerestory windows that run around the perimeter of the ground floor and intermittently on the upper level. The museum establishes its own context within the confines of the site itself, separated from all of its neighbors by an encircling swath of lawn.

Inside, the museum is dimly lit; the focus of attention is on a series of twenty-four paintings depicting scenes from Guru Tegh Bahadur's life, executed by artists Kirpal Singh (1923–90) and Jaswant Singh (1918–91).[17] The paintings are placed on both levels, and a visitor can follow events in the Guru's life in loosely chronological order by walking in a clockwise direction, starting on the lower floor and moving upward. Art historian Umri Kessar has described Kirpal Singh's paintings as focusing on events that were "narrative and didactic, seeking to exemplify the valour, the egalitarianism, [and] the magnanimity of the Sikh faith."[18] The paintings of Jaswant Singh, on the other hand, are saturated with graphic depictions of dismemberment, torture, and suffering at the hands of the Sikhs' Mughal enemies. His painting of Guru Gobind Singh leaving Anandpur along with followers at night, for example, places considerable demands on a viewer's eyes; one has to work to make things out on the dark canvas in the darkened hall. The emotional tone of the scenes depicted and the style of sensory engagement they demand work in unison to personalize the viewer's relation to the events as a kind of somatic experience. The quietude of the building's interior aids this effect by focusing the viewer's attention squarely on the visual narrative.

5. Main entry of the Guru Tegh Bahadur Museum in Anandpur Sahib, Punjab, India, 2008. Photograph by author.

Despite what one might expect to find in a modern "museum," the Guru Tegh Bahadur Museum, by narrowing and controlling the architectural context such that the visitor's empathy for the Guru can be secured as a visceral experience, foregrounds the continuing presence of the Guru—literally as a quality within the viewer's own body—rather than the "pastness" of the Guru. One should quickly add that it would be a mistake to see this museum as simply pandering to unsophisticated Sikhs who, in the form of a *kar sevak,* might unwittingly do damage to their own "heritage." This is at least one of the resonances *kar seva* activity has had for critics who uphold the value of empirical evidence over emotion in assessing historical value or, in slightly different terms, for those who uphold history over heritage, the latter of which is by definition concerned with the everpresentness of the past.

A second, more recent building in Anandpur figures the Sikh past in ways that both coalesce with and differ from this strategy, and help us further complicate the question of what roles heritage and history currently play at historical Sikh sites. The Khalsa Heritage Complex was designed by architect Moshe Safdie (who reportedly has Israeli, U.S., and Canadian citizenship) to commemorate the tercentennial celebration of the founding of the Khalsa by Guru Gobind Singh in 1699.[19] The Heritage Complex was initiated by Punjab chief minister Prakash Singh Badal, who identified Safdie following a visit to Jerusalem in 1997, where he encountered Safdie's Holocaust Memorial Museum. The Heritage Complex is not without its critics. Critics point out that Badal did almost no research on the project prior to hiring Safdie in a noncompetitive, private arrangement. Others criticize Safdie's design fee, which is rumored to be more than US$20 million, an amount that many argue would go a long way toward improving existing Sikh properties in the Punjab. Other complaints include the amount of resources the completed complex would consume, from water and electricity—the exhibits planned for the complex are based on laser and fiber-optic technology, huge plasma screens for the exhibitions, etc.—to land, money (reputedly more than $60 million for the building and exhibition), and even the number of historic buildings that had to be demolished to improve site access.[20] Still others have criticized the project because the designer is non-Sikh, non-Indian, Jewish, North American, and so forth.[21] Setting these concerns aside for the moment, let us briefly consider the Khalsa Heritage Complex in light of the discussion thus far.

The first thing to note about the building is that it shares many features with the earlier museum for Guru Tegh Bahadur. Both buildings are architecturally ambitious, abstract, "up-to-date," and predicated on establishing their immediate context within the confines of their own site rather than relating themselves formally to their built surroundings. The Khalsa Heritage Complex comprises two groups of buildings separated from one another by a 165-meter-long bridge that spans a series of reflecting pools. The western part of the complex is where visitors enter the site and will contain a "research and reference library" with archival materials, books, journals and audiovisual resources. The eastern part of the complex comprises permanent exhibit galleries and high-tech displays. While each of the architectural elements of the complex have been coded by the designer as symbolic of something integral to Sikh faith and history—the tower-like prismatic shafts that make up the main exhibition spaces of the museum wing are five in number and have been related, for example, to the idea of the *panj piyari* [five friends, or companions, of the Guru]—we can dismiss outright much of this as clever packaging: the Khalsa Heritage Complex doesn't so much accord with Sikh iconography and symbolism as arrest and suspend that iconography in exchange for global "brand" recognition—the Khalsa Heritage Complex is recognizably the signature design of an elite international architect, and the appeal of that "brand" is what led Badal to Safdie in the first place.

6. The Khalsa Heritage Complex, designed by Moshe Safdie, under construction in Anandpur Sahib, Punjab, India, in May 2007. The structures on the right side of the photo comprise the main exhibition galleries. To the left is the bridge connecting the eastern and western halves of the complex. Photograph by author.

If the Khalsa Heritage Complex has a formal strategy in common with the Guru Tegh Bahadur Museum, then it also partly shares the latter's preference for somatic engagement. The official website of the Heritage Complex describes the Museum as "telling a story" that's "deeply spiritual, passionate, and emotional." The exhibitions are organized as a "multi-layered communication strategy" whose goal is to create an "immersive" environment capable of "transport[ing] visitors into a different time and space, thereby enhancing their capacity to receive the intended communication. . . . The visitor to the Khalsa Heritage Museum," according to its website, "will not only leave better informed but will also be emotionally moved."[22]

Visitors begin on the second level, moving through a series of exhibits about the lives and teachings of the ten Gurus, where important events in the history of the Khalsa are delivered theatrically through "larger than life" sound, moving pictures, and dramatic lighting effects. The visitor then moves down along a spiral ramp to the lower level where "galleries chronicle Khalsa trials, tribulations, and triumphs from Banda Bahadur up to the immediate aftermath of Partition, when Sikh dynamism transformed the Punjab with its élan, energy, and resilience."[23] The route culminates in a serene reflective space, where a shaft of light that has been intermittently visible as the viewer circulated through the exhibits terminates in a lozenge-shaped patch of earth.

The array of high-tech media deployed at the Khalsa Heritage Museum differs only in kind from the graphic paintings at Guru Tegh Bahadur's museum; both share far more in terms of their intended effects than they differ from one another in the way those effects are delivered. Both attempt to draw the visitor bodily into an engagement with the past such as to make the past literally "come alive." What is interesting about Safdie's complex however is that "history" also gets its due—the archive, library, and reading room in the western precinct of the complex should appeal to those who, like the preservationists we encountered earlier, uphold the value of academic research as a guarantor of historical legitimacy. In other words, Safdie's building complex captures both ends of the doubleness of Sikh historical imagination, while proffering a hybrid new building format—the Heritage Complex—one well-attuned to the necessities of international heritage tourism (an explicit goal of the project).

How this story unfolds in the near future will tell us something else important about Sikh architectural practices and about the historical sensibilities they are called on to annotate. The understandable lament that Sikhism's original old buildings are no longer available for us to understand what Sikh architecture "really" is can only be a starting point, however, for broaching this larger question. The processes of rebuilding older structures and building new ones reveals something equally important about what Sikh architecture may or may not be. While preservationists deserve a prominent place in the conversations and political processes through which Sikh historical landscapes will be cared for in the future, the incomplete record of historical monuments need not be seen as an insurmountable obstacle for enriching knowledge about the Sikh past.

Biography

William J. Glover is associate professor of architectural history and theory at the University of Michigan. He is the author of *Making Lahore Modern: Constructing and Imagining a Colonial City* (Minnesota, 2008) and several articles on the intersection of built environments, knowledge cultures, and urban processes in colonial South Asia. Glover's current research is directed toward understanding the aesthetics of modernization in mid-twentieth-century South Asia, with a particular focus on new urban environments designed to receive rural immigrants, or, in contemporary sociological parlance, "villagers in the city."

Notes

I would like to gratefully acknowledge the financial support I received from the Bhai Gurdas Fellowship at the University of California at Santa Barbara, and acknowledge in particular the encouragement and insight offered to me by professor Gurinder Singh Mann. I also benefited from the comments of an anonymous reader of the draft manuscript of this article, which were very helpful in clarifying and strengthening the work. Any errors of fact or judgment that remain in the piece are mine alone.

[1] James Fergusson, *History of Indian and Eastern Architecture*, vol. 2 (New York: Dodd, Mead and Company, 1899), vi.

[2] Ibid., 83–84; emphasis added.

[3] This article is focused primarily on gurdwara rebuilding activity in India, given my focus on sites whose historical importance dates back several hundred years. There are many historically important sites in Pakistan that fit this criterion as well, of course, though few of these have been subjected to the vigorous dynamics of rebuilding described below. Many important historical sites in Pakistan were apparently rebuilt earlier in the twentieth century, but precious little research has been done on this topic. I am also leaving out the thousands of gurdwaras built around

the world by expatriate communities, many of which would richly merit being described as "magnificent."

[4] Among others, I have found the following works relevant to the issues explored here: Raphael Samuel, *Theatres of Memory, Volume 1: Past and Present in Contemporary Culture* (London: Verso, 1994); Valentine Daniel, *Charred Lullabies: Chapters in an Anthropography of Violence* (Princeton, N.J.: Princeton University Press, 1996); Michael Herzfeld, *A Place in History: Social and Monumental Time in a Cretan Town* (Princeton, N.J.: Princeton University Press, 1991).

[5] See, for example, Alois Reigl, "The Modern Cult of Monuments: Its Character and Origin," *Oppositions* 25 (Fall 1982): 21–51.

[6] Daniel, *Charred Lullabies,* 28.

[7] Gurmeet Rai and Kavita Singh, "Brick by Sacred Brick: Architectural Projects of Guru Arjan and Guru Harbobind," in *New Insights into Sikh Art,* ed. Kavita Singh (Mumbai: Marg Publications, 2003), 33.

[8] Gurmeet Rai quoted in Chander Suta Dogra, "Have You the Eyes for It?" *Outlook India,* May 5, 2008: http://www.outlookindia.com/article.aspx?237371.

[9] Patwant Singh and Gurmeet S. Rai, "Marble and Memory," *Indian Express,* December 23, 1998: http://www.indianexpress.com/res/web/ple/ie/daily/19981223/35750424.html.

[10] Ibid.

[11] Pardeep Singh Arshi, *Sikh Architecture in Punjab* (New Delhi: Intellectual Publishing House, 1986), see Figure 1, n.p.

[12] Demolition and reconstruction work at the Hazur Sahib gurdwara in Nanded, Maharashtra, is perhaps the most vociferously criticized project in the recent past. See, for example, Gagandeep Ahuja, "Leading Heritage Conservator Seeks Sack of Psricha from Hazur Sahib Committee," *Punjab Newsline,* January 24, 2007: http://www.punjabheritage.org/architectural-heritage/leading-heritage-conservator-seeks-sack-of-pasricha-from-hazoor-sahib-committee.html.

[13] Gurinder Singh Mann, "Sources for the Study of Guru Gobind Singh's Life and Times," *Journal of Punjab Studies* 15, no. 1–2 (2009): 6.

[14] Anne Murphy, "History in the Sikh Past," *History and Theory* 46, no. 3 (October 2007): 360.

[15] Public museums were first constructed in the 1840s in India and were common by the turn of the twentieth century in all major cities (and many minor ones). See Savithri Preetha Nair, "Economic Logic versus Enlightenment Rationality: Evolution of the Museum–Zoo–Garden Complex and the Modern Indian City, 1843–1900" in *Museum Revolutions: How Museums Change and Are Changed,* ed. Simon J. Knell, Suzanne MacLeod, and Sheila Watson, 61–71 (Abingdon, U.K.: Routledge, 2007); Tapati Guha-Thakurta, *Monuments, Objects, Histories: Institutions of Art in Colonial and Postcolonial India* (New York: Columbia University Press, 2004).

[16] See Reeta Grewal, "Three Centuries of Anandpur: The Town Where the Khalsa Was Instituted," *Journal of Punjab Studies* 15, no. 1–2 (2009): 65–93.

[17] A plaque at the entry to the museum lists the artists as S. Kirpal Singh, S. Jaswant Singh, and S. Devinder Singh.

[18] Umri Kessar, "Twentieth-Century Sikh Painting: The Presence of the Past," in *New Insights into Sikh Art,* ed. Kavita Singh (Mumbai: Marg Publications, 2003), 125.

[19] The client for the complex is the Anandpur Sahib Foundation. The Indian associate architect for the project is New Delhi–based Ashok Dhawan.

[20] These and other controversies are discussed in Anne-Colombe Launois (Sat Kaur), "The Khalsa Heritage Complex: A Museum for a Community?" in *New Insights Into Sikh Art,* ed. Kavita Singh, 134–45. On the political context within which Badal's scheme emerged, see Praveen Swami, "A Fusion of Politics and Religion," in *Frontline* 16, no. 9 (April 24–May 7, 1999). Available at http://www.frontlineonnet.com/fl1609/16090350.htm.

[21] See, for example, Singh and Rai, "Marble and Memory."

[22] From the official Khalsa Heritage Complex website: http://khalsaheritagecomplex.org/khc.html.

[23] Note how the latter comment assimilates Sikhs into a broadly nationalist frame through a narrative that elides post-Partition demands by Sikhs for a separate homeland and emphasizes instead their contributions to the nation-state. See Harjot Oberoi, "From Punjab to 'Khalistan': Territoriality and Meta-commentary," *Pacific Affairs* 60, no. 1 (Spring 1987): 26–41.

1. Yunti street in Xituo. Photograph by author.

Paola Demattè

After the Flood

Cultural Heritage and Cultural Politics in Chongqing Municipality and the Three Gorges Areas, China

The Three Gorges Dam: The Strategy of Limited Listening

The construction of the world's largest hydroelectric dam on the Yangzi River has caused unprecedented transformation in southwestern China. By halting the course of the Yangzi River in proximity of the scenic Three Gorges area, the dam has led to the creation of an artificial reservoir that covers more than one thousand square kilometers. Over the years, the rising waters have flooded thirteen cities, 140 towns, 1,352 villages and more than one hundred thousand acres of agricultural land, and have required the resettlement of 1.3 to 1.6 million people.[1]

Although, ostensibly, it was seen to bring great benefits to the country (energy, flood control, and increased navigation to China's interior), the Three Gorges Dam project has long faced controversy and contestation.[2] The idea of a dam blocking the Yangzi at the Three Gorges was first proposed early in the twentieth century by Sun Yat-sen, the founding leader of the Republic of China, and reproposed by Mao Zedong in the 1950s. However, given its catastrophic potential for environmental destruction and social displacement, the project always had opponents both abroad and within China.[3] As documented by Dai Qing, a prominent journalist and human rights activist, the approval of the project by China's National People's Congress, which took place between 1989 and 1992, encountered dissent from various quarters.[4] Sociologists, cultural heritage specialists, and rights activists, who like Dai Qing were at the forefront of the fight against the Three Gorges Dam, were the most vocal in this resistance. Their voices were also the first muzzled. Dai Qing herself was imprisoned following the June 1989 student protests, because, according to her critics, *Yangtze! Yangtze!*, her 1988 report on the Three Gorges project, was responsible for causing "riots and chaos." Beyond these notable critics, opposition to the project was wide and included many technical experts as well as local and national leaders. A 1995 report by Human Rights Watch/Asia indicates that in the April 1992 vote to approve the "Resolution on the Construction of the Yangtze River Three Gorges Project," one third of the delegates of the National People's Congress either

Future Anterior
Volume IX, Number 1
Summer 2012

abstained or voted against it.[5] Heritage experts and officials of the State Administration for Cultural Heritage (Guojia Wenwuju) at the local and national level also expressed dissent and concern at the scale of the destruction. However, the institutions pushing the project trumped the available venues of critique, which were limited to begin with, since the political system is controlled by the same entity that supported the project, the Chinese Communist Party. Though easily theorizable on a supranational scale as repressive strategies, these institutional pressures are better understood within the cultural contexts and the historical constraints of those who exerted as well as experienced them. First, the project was desired by then-Premier Li Peng, who as a hydraulic engineer–turned-politician saw the dam as the crowning of both his professional and political careers.[6] Second, the dam was presented as a project that was not only necessary for modernization but also as a source of national pride, particularly in the face of international criticism of the project. "Biggest and best in the world" were terms that circulated both in the national and international media, but in China they were perceived as code for "patriotism." As we have seen in the American media recently, critics of projects that are a source of "national pride" or concern for "national security" can be effortlessly framed as unpatriotic self-haters. Similarly, Chinese citizens presented with these overt and oblique narratives could easily interpret their underlying implications. For millennia, Chinese have experienced the state as a giant bureaucratic machine that moves slowly but inevitably toward goals that supposedly trump individual or local rights for an ambiguously defined "greater social good," a concept that the Chinese government has termed "Confucian human rights."

Eventually, against this symbolic pressure, a sense of inevitability set in. The debate turned toward making the "best out of the difficult circumstances" and saving the salvageable. The sense of inevitability, however, was not just the result of bureaucratic rigidity or political oppression; it was an outcome of subtle communication strategies aimed at responding to different constituencies. The resulting form of the project and its concomitant devastation occurred not only because criticism fell on the state's deaf ears but because its impact was mitigated by an ambiguous pretense of listening and responding. This strategy—of appearing to respond to preservation demands but only in a way that didn't restrict in any way the fundamentals of the dam project—succeeded in portraying the project as a participatory and culturally sensitive enterprise in the national media. The cooperation was presented as a unifying, cooperative effort on behalf of the nation.

Chongqing and Its New Municipality: Absorbing Displacements

Though the dam body was built at Sandouping, near Yichang (Hubei province), the Three Gorges Reservoir, which stretches for 640 km (385 miles) to Chongqing city, is for the most part in Chongqing's territory. The greatest impact of the dam project was therefore on the municipality of Chongqing, which was created in tandem with the development of the dam.

In 1997, the city of Chongqing (formerly known as Chungking) and surrounding counties affected by the dam project were granted autonomy from Sichuan province and elevated to the rank of direct-controlled municipality, like the metropolises of Beijing, Shanghai, and Tianjin. With an area of 82,300 km^2 and a population of approximately 32 million, including six million urban residents, Chongqing is now one of the world's largest single metropolitan areas. Beyond Chongqing city (the administrative center is home to 2.5 million people), the municipality includes nineteen districts with sizable cities like Wanzhou (300,000) and Fuling (200,000), and twenty-one counties, four of them autonomous and inhabited by ethnic minorities. The creation of the municipality was likely a trade-off that the central government (which was the owner of the dam project) struck with the regional political entities. But by offering a degree of autonomy and funds to improve the local economy, Beijing was able to bypass local opposition and obtained the political support needed to proceed with implementation. In fact, the project was never controlled by the local governments of the areas affected by the dam (which were two: Chongqing municipality and Hubei province) but directly from the center. At the same time, the creation of the Chongqing municipality was a means to facilitate the project that could also be presented to the local inhabitants as a source of identity pride.

Most of the territory of Chongqing municipality was affected by the Three Gorges project. The areas that were to be flooded were progressively destroyed, and the areas not touched by the flooding underwent major transformations to make space for massive relocations and developments. In the late 1990s, when the dam was under construction, cities and villages throughout the future reservoir were haunted by ominous red signs that indicated the yearly projected levels of the water as time progressed (especially ominous were the marks for 175 meters, which was the projected water level by the completion of the project in 2009). Although it is standard procedure in dam construction to mark the progressive flooding levels on site so that infrastructure (buildings, sewage, utility lines etc.) can be dismantled and removed prior to the arrival

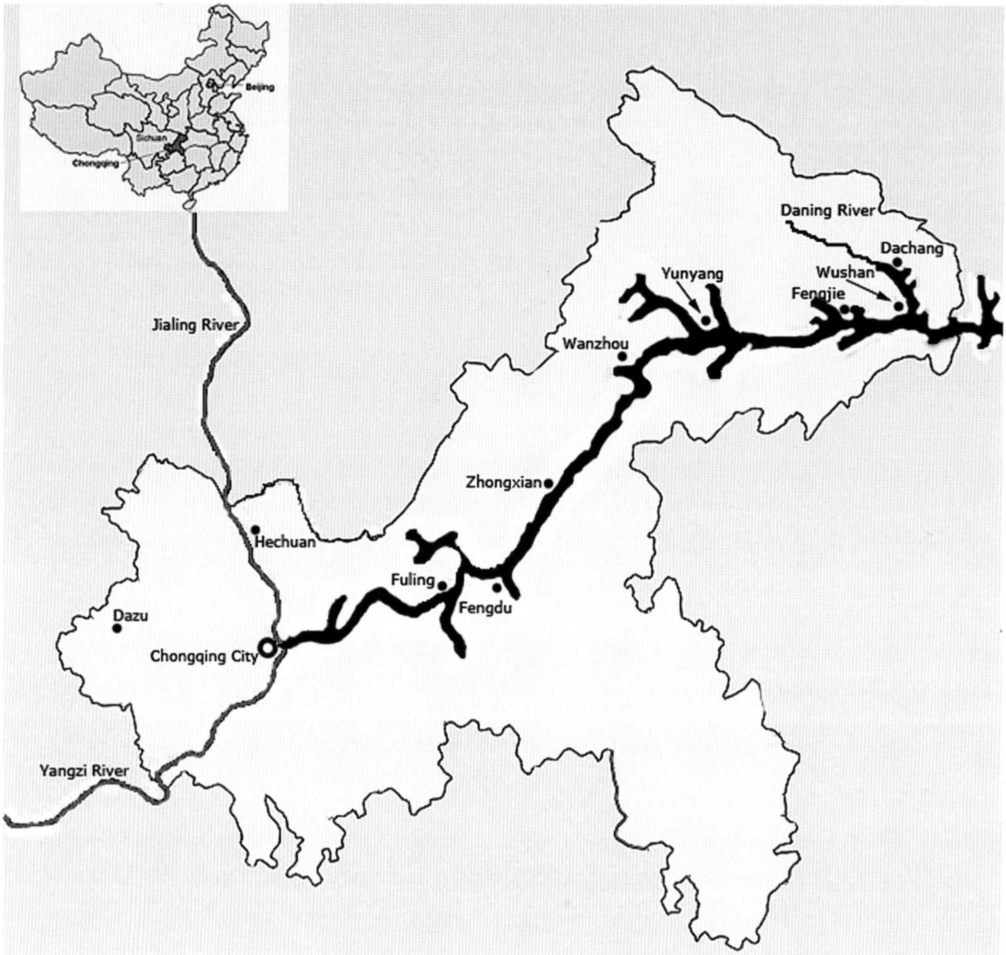

2. Map of Chongqing Municipality. Map by author.

of the waters, it is clear that in the Three Gorges (as elsewhere) this signage was also meant to intimidate the residents: accept the relocation payments and get out now, because by *X* year your house, your land, and your tombs will be under water. Residents in some cases resisted the idea that the flooding of their property was going to be real, but for the government there was no space for unpredictability and negotiation. However, if dam construction and flooding proceeded rigorously as planned, steered by Three Gorges Dam Construction Committee under the direction of the fine-tuned Chinese state machine, the same cannot be said about relocations, compensations, and preservation. Local corruption often short-changed the most vulnerable inhabitants, who found themselves dispossessed and displaced. Similarly, preservation was considered secondary to the urgency of construction, which in any case was more lucrative. If the progressively rising water level created a feeling of looming doom in the settlements that were about to be flooded, just above the flood line this anguish was countered by a frenzy of construction that, with its large-scale transformations, became in turn further justification for the

3. Flooding mark at Fuling. Photograph by author.

creation of Chongqing municipality. Though the creation of Chongqing was presented as a remedy to the impact of the Three Gorges project, the frenzied pace of construction that it generated set in motion a consensus of inevitability, leaving no room to question how the new development transformed and eradicated what was being displaced.

The construction of the dam has undoubtedly brought state-funded economic growth to this remote area of China, but the impact of the project has been devastating for the displaced rural inhabitants, the environment, as well as the material and intangible heritage. Though postmodern interpretations question the notions of "cultural heritage," "local culture," or "traditional culture" as essentializing categories that do not exist outside of history unencumbered by various politico-economic influences, the inhabitants of the Three Gorges area perceived that the realities of "their lives" and "their places" were dissipated by a powerful but distant entity, which, unlike in the past, they were in no position to question. As archaeological and architectural sites both below and above the final flooding level of the reservoir were lost or endangered, old economies and social setups were critically disrupted or simply extinguished. These new challenges added to and exacerbated preexisting physical threats—such as industrial pollution (with resulting acid rains and saltification of ground water), excessive urban development, and looting—to small rural settlements and to cities. The changes brought about by the dam construction drastically transformed the previous checks and balances on resource consumption and exercise of power, and rewrote them on an unusually large scale, cutting the citizens out of the debate. The new transformations introduced a pace of change and unbridled growth that created new and extensive networks of political corruption, careless tourist exploitation, and problematic uses and interpretations of the past and of people, which in more insidious ways precluded the possibility of organization and resistance by those who suffered their effects.

Chongqing: The Question of Archaeological and Historic Heritage

The Three Gorges Dam's construction and the associated creation of Chongqing municipality redefined the ideas of "local heritage" and "local culture." In its effort to create a local identity to go with the new administrative structure, the new municipality encouraged the Cultural Bureau and subsidiary cultural institutions such as museums, research institutes, and universities throughout its territory to think in terms of a shared local identity that was to be sponsored and supported to make both residents and tourists feel included in the project. Although the politics of "local heritage" existed before the establishment of the municipality, afterward the municipality became the locus and the reason of central identity management.

Historically, Chongqing had a rich and idiosyncratic material and cultural heritage. The area included unique living cultures as well as archaeological and historic sites ranging from the Paleolithic to the contemporary.[7] This heritage was and in part still is concentrated in the vicinity of the Yangzi river watershed—particularly in the city of Chongqing and in the Three Gorges area—but a peculiar aspect of its intangible cultural, social, and economic practices is found in other parts of the municipality as well, such as Qianjiang autonomous county, home to the Tujia and Miao (Hmong) ethnic minorities.

The Three Gorges was a self-contained area that for millennia was defined by the flow of Yangzi River and its tributaries. One of its characteristics was the riverine cities, villages, and small towns that with their vernacular architecture created a complex river network. Perched on the steep sides of the Yangzi valley, these settlements featured the typical fluvial urban fabric, their gates opening onto the river with vertical stairs leading down to the water.[8] A fine example of these structures was Kaiji gate at Fengjie (now submerged). Another was at Xituo, a village in the Shizhu Tujia Minority autonomous county, which though submerged in its lower section, lives on as a tourist attraction. In antiquity, the village was defined by the stairs of Yunti Street that snaked from the top of the settlement to the river wharf that connected with the outside world.

4. Kaiji gate at Fengjie. Photograph by author.

Villages and towns on the tributaries of the Yangzi River were also part of this system. In its preflood days, Dachang, a 1,700-year-old small town in the mid–Daning River Valley in Wushan county, featured well-preserved Ming (1368–1644) and Qing (1644–1911) city walls and gates, aristocratic residences, and temples. Dachang was famous for its intimate connection with the river. Before the introduction of motorboats, its inhabitants built a series of wooden passageways on the sides of the river to pull their boats up from the Yangzi. In antiquity, the connection with the river was maintained even at death and people were buried over the Daning in coffins hanging from its steep cliff sides.[9]

The Three Gorges and its network of towns were also symbolically identified by a system of landmarks that dotted the river at various points and served as signs marking the different points of the river. The best known are the ancient hydrometric station of Baiheliang (Fuling), the Ming period complex of Shibaozhai (Zhongxian), Zhangfei Temple (Yunyang), and the Baidicheng citadel (Fengjie). Most of these historical and religious structures are associated with river lore or with historical legends relating to the Three Kingdoms period (220–280 CE), and though dating mostly to the late imperial period (Ming-Qing), they are developments of older structures and settlements.

Also related to this riverine network were several ancient military garrisons and fortified citadels, like Diaoyucheng (near Hechuan, on the Jialiang river), Panshicheng (Yunyang), and Tianshengcheng (Wanzhou), which occupied the northern

5. A street in the old village of Dachang. Photograph by author.

shore of the river. These defensive structures, which survived until recently in various states of preservation, are well known for the role they played during the southern Song period (1127–1279), when they were employed in conjunction with the Yangzi river in the effort to stop the Mongols from invading southern China. In some cases, their history stretches back as much as two thousand years and their use has continued through more recent centuries, expanding to include religious activities and rituals.[10]

The flooding obliterated many of these material and immaterial networks, but the destruction was not the only problem; this network of associations was eradicated also by the politics of the newly established Chongqing municipality that envisioned preservation as the relocation of select landmarks for museum-like exposition. These preservation and relocation projects effectively disconnected these sites from the geography that created and sustained them. The isolated examples of the physical fabric "preserved" and transported to new locations reflected more the need to create monolithic and nonoverlapping cultural identities that could be used for political maneuvering and tourist consumption and not the sociophysical context submerged by the waters.

A critical example of this cultural redefinition is the ongoing reframing of the Ba culture. Among the many archaeological and paleontological sites that have been submerged by the dam, archeologists and cultural historians have paid particular attention to assemblages dating to the Neolithic and Bronze Ages, which are said to belong to the so-called Ba culture (circa 2000–200 BCE), a horizon documented by preliminary excavation at Shuangyangtang and Lijiaba, as well as at Ganjinggou, Xiaotianxi, and Zhongba.[11] Though still little known, Ba was a culture that, while distinct in many respects, was intertwined with surrounding cultural practices and communities that occupied the Yangzi River watershed now partially submerged by the Three Gorges Reservoir.

In the years leading up to the flooding, Ba culture was deployed as an isolated and distinct heritage to justify selective relocation enterprises, even though, strictly speaking, Ba refers only to a late Neolithic and early Bronze Age phase. Nonetheless, Ba played an increasingly important role in the historical self-definition of Chongqing municipality, particularly as a form of distinction from Sichuan province, its larger neighbor and former administrative center, which is considered to be the center of a different culture, Shu. However, when seen in their larger intertwined historical setting, the differences between the two prehistoric cultures and their historic descendents appear not as marked as local authorities make them out to be. Nevertheless, Ba culture became the ideological motor and uniting factor behind many cultural enterprises, from the preservation and marketing of the material heritage, to the construction of museums and cultural theme parks, to the coopting of Tujia minority peoples as living descendants of the ancient Ba.[12] In short, Ba became a general narrative that united the local heritage and guided the selection and relocation of material heritage to designated culture areas.

Chongqing: Planning Preservation

If some of this selection was intentional, driven by demands for cultural definition and to some degree by tourist interests, the rest was the consequence of the pace of the project and its presumed inevitability. Prior to the beginning of the dam project, it was estimated that at least 1,238 archaeological sites and historic monuments in the Three Gorges and surrounding areas were going to be affected by the flooding. Of these, 845 were in Chongqing territory, and the remaining in Hubei.[13] Among these, two-thirds were archaeological sites predating the thirteenth century, while the remaining were above-ground architectural landmarks dating to the late imperial phase (Ming and Qing).[14] In the 1990s, government experts surveyed the then-projected reservoir area, drawing protections plans for more than sixty aboveground historic sites and setting up archaeological investigations for five hundred others. The declared aim of the survey was to preserve the most important elements of the "local cultural heritage." However, this goal was already reductive, missing the multidimensional connections that tied these "important landmarks" to an array of immaterial practices and particular geographical settings. As the preservation plans developed, even this limited goal was reduced in size and scope.[15]

Archaeological work began in earnest only in 1997 and at a pace that would have required thirty years to complete investigation of all the sites identified. The problems facing the proposal were many. First, Three Gorges archaeology was a big unknown before the salvage work began. Second, once the importance of the area emerged, it was clear there was neither enough qualified manpower in China nor enough funding to address this emergency. Third, the Dam Construction Committee was unwilling to accept foreign funding or foreign help in the archaeological investigations because, after widespread international criticism, the Three Gorges project became a patriotic show of China's growing power. As a result of these policies, many sites have remained unexcavated and are now lost to researchers, whereas others have been destroyed by the necessities of a hasty development. However, this is not to say that the Chinese archaeological community easily succumbed to the sense of inevitability that the masterminds of the project were projecting. Preservation or excavation of all sites is never feasible or desirable anywhere, but archaeologists from all Chinese universities united in a massive excavation effort and obtained a significant amount of data, even though they were under enormous pressure.

Concerning the aboveground cultural heritage, such as historic temples and vernacular architecture, the problems were even more serious. Due to economic and technical problems, only a limited number of sites that were to be completely or partially flooded could be protected *in loco* or relocated to higher grounds. Finally, protection plans were implemented for three sites in Chongqing territory: Baiheliang (Fuling), Shibaozhai (Zhongxian), and Zhangfei Temple (Yunyang). The most unusual was the construction in Fuling city of an underwater museum for the *in situ* protection of the ancient hydrometric station of Baiheliang, a rock outcrop that bore the historic inscriptions documenting the Yangzi River water levels and floods.[16] In similar fashion, the rock of Shibaozhai (Zhongxian) was protected in its original location with the construction of a coffer dam. Thus encircled, Shibaozhai, a thirteen-story, fifty-meter-high Ming temple complex built on a cliff overlooking the village of Shibao on the northern shore of the Yangzi River, has become an island in the midst of the reservoir. Finally, Zhangfei Temple, a structure dedicated to third-century local hero General Zhangfei, was relocated thirty kilometers upstream at the village of Long'an, opposite the new town of Yunyang. For most of the remaining heritage, only partial relocation was considered and in many instances nothing was saved.

6. The Baiheliang river station at Fuling. Photograph by author.

These three examples show how the official response was two-pronged. On the one hand it focused on the preservation of famous landmarks, and on the other it promoted the idea of the "saved" landmark as a museum experience. This fascination with the museum is not limited to the Three Gorges area but is widespread throughout China. Officials like the idea of supporting institutions such as museums, not only because culture is valued in China but also because, more than an ancient building in its original setting, a museum or a cultural park is a place of entertainment that brings tourist profits.

The problems of the Three Gorges heritage are not limited to those sites that were actually flooded by the waters. Many other sites that were either partially or well above the flooding level are in danger as well, either because of rising waters nearby or due to new urban development. Since allocations for the protection and/or excavation of cultural heritage were limited by the Dam Construction Committee to only those sites directly affected by the new reservoir, anything above the 175-meter level was technically excluded from the protection plans and budgets. With no provisions for their protection, many face various challenges and may ultimately disappear.

For instance, the Neolithic site of Daxi (Wushan) and the historic Baidicheng complex are only partially affected by the waters that have flooded their lowest portions, but the higher water table and the ingress of pollutants threaten their stability. Other sites, though protected, are indirectly affected by the dam because relocated cities are too close or encroach on their environment. The Song citadel of Tianshengcheng (Wanzhou), known for its ancient walls and historic calligraphies, is endangered by the resettlement plan for Wanzhou, which brings the new city close to the site and has already led to the construction of a cable car station atop the ancient walls.

Sometimes relocated cities are actually within archaeological sites, as at Wushan, where the new city was built in the midst of Paleolithic and Neolithic sites and a Han dynasty (206 BCE–220 CE) cemetery. In this case, some fieldwork was carried out, but often archaeological sites uncovered during construction were looted and destroyed, because fast-paced construction and time constraints made supervision lax. In the preflood years, it was not uncommon to see archaeological items for sale in the markets of the towns to be relocated. Artifacts illegally excavated in the area were also sold abroad with the complicity of construction officials.[17] This destructive but lucrative trade (which is by no means unique to the Three Gorges) was in defiance of existing laws dedicated to protecting the "national heritage" promoted by government institutions such as the State Administration for Cultural Heritage *(Guojia Wenwuju)*. In practice, however, the State Administration for Cultural Heritage is often powerless in the face of powerful political and business lobbies, fast-paced developers and the lucrative tourist industry and is rarely able to enforce these laws.[18]

While obliteration of cultural heritage networks and practices is much larger in the Three Gorges area, similar problems are also confronted in other parts of the Chongqing territory, which as a result of a cascade of investments related to the dam project, underwent major economic development. Important Buddhist and Taoist stone carvings, grottoes, and temples, which are quite common in the rural areas, are often threatened by religious and tourist development as well as by pollution and looting.[19] Historic structures in the city of Chongqing have also suffered considerably from development in recent years. Economic and political interests related to urban development have eaten up much of the city's existing historical architecture.[20]

Analysing the Situation

The logic behind the preservation strategy for both the material and intangible heritage of the Three Gorges area is sometimes hard to grasp. One of the main problems is the draining of intellectual energy and financial resources by grandiose and expensive "preservation plans" for the few high-profile structures discussed above (i.e., Baiheliang, Shibaozhai, and Zhangfei Temple). Due to their technical difficulty and great cost, these projects came at the expense of the preservation of supposed lesser sites and of the vital wider sociocultural context in which they were embedded. So, as few famous landmarks were very publicly "saved" from the flood, there was no definite salvage plan for the hundreds of villages and small towns that with their unique urban layouts and vernacular architecture made up the heart of the Yangzi River life. For the most part, these settlements disappeared quietly and completely. In the case of a few historic towns, a limited preservation plan was implemented that involved the relocation of some structures either to an open-air architecture museum or to a new site designed for tourist consumption. Far from ideal, these partial relocations have led to the creation of what can be called "historical fakes." An example of this predicament is Dachang, the seventeenth-century small town on the Daning River described earlier. The plan for the salvage of Dachang consisted of relocating a few of its best structures (such as the Ming-period Wenjia Da Yuan) in what appears to be a culture theme park five kilometers from the original site. What could not be moved was abandoned to the waters, and a replica was rebuilt at the new site. Although some part of the population moved with the new town, taking up the position of performers in the reconstruction of their former village, most inhabitants dispersed to different parts of China. What is most ironic is that those living at the new Dachang are now experiencing even more acute economic hardships. The new town seems to lack the appeal of the original and is visited far less by tourists.

Issues and Concerns

Many of the "preservation" strategies adopted to deal with the emergency in the Three Gorges Dam project are a source of major ethical concern, particularly in terms of how and by whom the sites that were saved will be used, interpreted, and understood in the near and distant future. From the onset, it was clear that the local inhabitants of these sites had little agency in this matter. The selective and arbitrary moves of historic structures led to an increasing lack of control over the framing of their heritage due to the distance or to the change in meaning of the relocated heritage. In all the cases described above, the "preserved" site is not only relocated at considerable distance or separated from its original users, but it is also completely transformed by the intervention (we can think about the hydrometric station of Baiheliang, which for centuries marked the water level of the Yangzi in the port of Fuling and is now completely underwater *and* in a museum). The sense of separation that these plans embody is reiterated at different scales throughout the reservoir area. It goes from the sense of loss of a community deprived of a landmark to the erasure of memory generated by the destruction of the fluvial network of riverine cities and villages.

In the end, what is left of the heritage is often just a distorted image of its original self, cleared of any meaning and rebranded for sale to a growing tourist audience. In fact, as the process of modernization and overall economic growth in China led to the destruction of countless historic sites, the well-off have developed a longing for the "past" and "tradition." Unfortunately, the past and traditions they have themselves pulled into existence, such as culture parks and megamuseums conceived by the tourist industry, are parodies of the historical realities they replace. In these transformed contexts, the "saved" heritage survives as a simulacrum of its former self.

For example, to respond to the identity crises faced by Chongqing and the Three Gorges and to boost growth and tourism, politicians and even scholars have resorted to the construction of symbolic culture landmarks meant to reiterate the historical identity of the area. One such creation is the Ba Culture Village, a park in the outskirts of Chongqing, which includes both modern buildings in "traditional" style and old monuments moved from flooded or newly developed areas. Another is the large museum dedicated to the Three Gorges culture that was built in the center of Chongqing to house recently excavated artifacts as well as relocated heritage.[21] Although the Three Gorges Museum is overall a positive development, it is hard to deny the feeling of generalized loss as more and more of the historical heritage is heading inside a box, away from the people that until recently still made use of it.

This is not an isolated problem in contemporary China. Here, as elsewhere in the world, preservation of antiquities is understood by most entrepreneurs and politicians as a means to further the exploitation of their tourist, economic, and political potential. While this is perfectly logical and to a certain degree understandable, this approach to cultural heritage often leads to damaging and arbitrary reconstructions of the past. As such, the danger for the cultural heritage is not only destruction but actually reconstruction, manipulation, and museification to suit local political and economic ends. It is difficult to say what is the right solution for the cultural and living heritage of Chongqing, the Three Gorges, and for that matter of all of China at this stage. Perhaps what is needed to avoid the museification and exploitation of material and living culture and peoples are educational campaigns aimed at political, business, and religious leaders but also at cultural heritage specialists and scholars.

Biography
Paola Demattè is associate professor of Chinese art and archaeology in the Department of Art History and Visual Culture at the Rhode Island School of Design in Providence. She holds a *Laurea* in Chinese language and literature from the Università degli Studi di Venezia and a PhD in archaeology from the University of California, Los Angeles. Dr. Demattè has curated an exhibition at the Getty Center and coauthored the award-winning *China on Paper* (2007), on Sino–European exchanges in the sixteenth through the nineteenth centuries.

Notes

[1] Yan Tan, *Resettlement in the Three Gorges Project* (Hong Kong: Hong Kong University Press, 2008). See also the review of this book by Jing Jun in *The China Journal* 60 (July 2008): 197–99, and his analysis of the resettlement problem: Jun Jing "Rural Resettlement: Past Lessons for the Three Gorges Project," *The China Journal* 38 (July 1997): 65–92.

[2] James Beattie, "Dam Building, Dissent, and Development: The Emergence of the Three Gorges Project," *New Zealand Journal of Asian Studies* 4, no. 1 (June 2002): 138–58.

[3] Shiu-hung Luk and Joseph Whitney, eds., *Megaproject: A Case Study of China's Three Gorges Project* (New York: Sharpe, 1993); A. R. Topping, "Ecological Roulette: Damming the Yangtze," *Foreign Affairs* 74, no. 5 (1995): 132–46; Jan Knippers Black, *The Politics of Human Rights Protection: Moving Intervention Upstream with Impact Assessment* (Lanham, Md.: Rowan and Littlefield, 2009).

[4] Dai Qing, ed., *The River Dragon Has Come! The Three Gorges Dam and the Fate of China's Yangtze River and Its People* (Armonk, N.Y.: M. E. Sharpe, 1998); Dai Qing, ed., *The Yangtze River: Arguments on the Three Gorges Project* (1988; republished by Cambridge, Mass: Fairbank Center for East Asian Research Library, Harvard University, 1991).

[5] Human Rights Watch/Asia, *The Three Gorges Dam in China: Forced Resettlement, Suppression of Dissent and Labor Rights Concerns* (New York: Human Rights Watch, 1995).

[6] Marta Ponseti and Jordi López-Pujol, "The Three Gorges Dam Project in China: History and Consequences," *Revista HMiC* 4 (2006): 151–88.

[7] *Jianghan Kaogu* 1 (1994), issue dedicated to Three Gorges archaeology.

[8] Deirdre Chetham, *Before the Deluge: The Vanishing World of the Yangtze's Three Gorges* (New York: Palgrave Macmillan, 2002).

[9] Yun Zhang and Cheng Zhu, "Environmental Archaeology of the Dachang Region in the Daning Valley, The Three Gorges Reservoir Region of the Yangtze River, China," *Chinese Science Bulletin* 53, sup. 1 (March 2008): 140–52; Eryu Luo, "Preliminary Study of the Sichuan Cliff Tombs," *Kaogu Xuebao* 2 (1988): 133–67.

[10] This network also extended beyond the Three Gorges area, to a large number of religious grotto sites, some Buddhist, others Taoist, and still others associated with local deities that covered the territory of what is now the Chongong municipality, including the Dazu grottoes, a celebrated World Heritage site.

[11] Steven F. Sage, *Ancient Sichuan and the Unification of China* (Albany: State University of New York, 1992): 47–82; Xu Zhongshu, ed., *Ba-Shu kaogu lunwenji* (Beijing: Wenwu, 1987).

[12] It is interesting to note that people of the Tujia minority have been enlisted in this project of Ba identity construction and are presented as the original inhabitants of the Three Gorges area and therefore the descendants of the ancient Ba people.

[13] Chongqing Cultural Heritage Bureau and Chongqing Settlement Bureau Chongqing, *San Xia wenwu baohu xueshu yantaohui lunwenji* (Beijing: Kexue, 2001); Three Gorges Dam Construction Committee and State Administration of Cultural Heritage, *Hubei kuqu kaogu baogaoji* (Beijing: Kexue, 2003).

[14] Elizabeth Childs-Johnson and Sullivan, "The Three Gorges Dam and the Fate of China's Southern Heritage," *Orientations* 7 (July–August 1996): 55–61; Dai Qing, "Interview with Yu Weichao," in *The River Dragon Has Come,* 124–42.

[15] Xu Guangji, Liu Yuchuan, Wu Hongtang, eds., *Yongbu shiluo de wenming: Sanxia wenwu qiangjiu jishi / Civilization Will Not Be Flooded: Documentation of the Rescue of Cultural Heritage in the Three Gorges* (Jinan: Shandong huabao, 2003).

[16] Ge, Xiurun, "Baiheliang Ancient Hydrologic Inscription: No. 1 Ancient Hydrometric Station in the World and In-situ Underwater Protection Project," http://www.unesco.org/new/en/culture/themes/underwater-cultural-heritage/about-the-heritage/underwater-museums/the-baiheliang-underwater-museum.

[17] E. Childs-Johnson, "Three Gorges Dam and the Looting of Archeological Treasures," published electronically in the International Rivers Network website: www.irn.org, May 1, 1998; Zhao Shilong, "Looters and Salvage Archaeology in the Three Gorges Reservoir Area," *Nanfang Zhoumo,* October 7, 1997, also published electronically in the International Rivers Network website: www.irn.org.

[18] Jeanette Greenfield, *The Return of Cultural Treasures* (Cambridge, U.K.: Cambridge University Press, 2007): 274–79, 357–62.

[19] Jiang Decai, "Thieves Who Hacked Away at Major Dazu Sculpture," *Zhongguo Wenwu Bao,* January 21, 1998: 1.

[20] One success story was the preservation of the Huguang Huiguan, a seventeeth-century former merchant-guild complex, which consisted of various structures including a courtyard theater. In the late 1990s, the Huguang Huiguan was an overcrowded rat-infested settlement in downtown Chongqing. Given its dilapidated condition and its location on prime real estate, the site was threatened by developers

who could capitalize on the high land value. It was brought back to some form of its original life with the combined support of the local- and national-level Administration of Cultural Heritage, the World Bank's Chongqing Urban Environment Project, and the Italian Trust Fund for Culture. The author was involved in the identification of this site for preservation while working as a consultant for the World Bank on the Chongqing Urban Environment Project between 1998 and 1999. See also Katrinka Ebbe, "Infrastructure and Heritage Conservation: Opportunities for Urban Revitalization and Economic Development," *Directions in Urban Development,* no. 48019 (February 2009): 1–4.

[21] Chongqing Museum, *Chongqing Zhongguo sanxia bowuguan* / China Three Gorges Museum, Chongqing (Beijing: Wenwu, 2005).

1. Ørnesvingen viewing platform, National Tourist Routes Geiranger-Trollstigen. 3RW Architects, Smedvig Landskap AS. Photograph by Simon Skreddernes.

Janike Kampevold Larsen

Global Tourism Practices as Living Heritage

Viewing the Norwegian Tourist Route Project

At the end of his 2010 novel *The Map and the Territory,* Michel Houellebecq describes Ruhrgebiet as a case of "industrial tourism." It is hardly a novel observation that the entire region is "like a conservatory of the first industrial age in Europe."[1] What the protagonist is particularly interested in, however, is "the menacing density of the forest that, after scarcely a century of inactivity, surrounded the factories." He observes that the preservation of the Ruhrgebiet, with its blend of dilapidated industrial complexes and terraced housing, allows for a certain renaturalization in the form of indigenous seeding and formation of scenery. What becomes negatively visible is the "perishable and transitory nature of any human industry." Houellebecq suggests here that contemporary landscapes are invented by tourism and rest on a rearticulation of preindustrial sceneries—uninterrupted and pleasing. But this is only after he has made manifest the fact that Chinese tourists are making the French "*terroir* " more authentic than it was—recognizing the need to return to local customs to please seekers of authenticity.[2] Authenticity is produced, both as scenery and custom.

An intriguing infrastructural scenery is currently unfolding in Norway. The Tourist Route Project (TRP), administered and developed by the National Public Roads Administration, displays landscapes invented by tourism.[3] It rests on roads built to accommodate the first tourists to the country in the late nineteenth century, and it produces layers of heritage, both tangible and intangible. This article discusses the TRP in relation to questions of heritage as defined by Gregory John Ashworth and J. E. Tunbridge as "the link between the preservation of the past for its intrinsic value, and as a resource for the modern community as a commercial activity."[4] It subsequently frames the TRP as a production of heritage and as a practice that accordingly problematizes the concept of *authenticity*[5]—all in the light of the global tourism industry. It explores the practice that structures the installations in the TRP and suggests that they perform an unwitting form of intangible heritage preservation by encouraging tourists to reenact viewing practices that are deeply founded in a nineteenth-century Norwegian—and European—tradition for viewing landscape. Thus the main dynamic is the construction of space and

Future Anterior
Volume IX, Number 1
Summer 2012

experience, which Susan S. Fainstein and David Gladstone define as integral to the authenticity of tourist-scapes, and the "contribution of tourism to urban and regional economies."[6] Also, the TRP's massive supply of panoramic and framed views certainly underscore Robert Silverstone's thesis that museums and heritage displays may be treated as mass media due to their "increasing appeal to the popular, and their common status as construction."[7]

The Tourist Route Project is establishing eighteen scenic byways throughout the country. The routes are not a continuous system but exploit sections of already existing roads, most of which were built between the 1880s and the 1940s. The most iconic routes are southern mountain passes, built to a large degree in order to bring tourists from the head of fjords up hillsides to the best viewing points. As an example, the Prussian Kaiser Wilhelm II visited Norway regularly between 1869 and 1912, before the country was at all prepared to accommodate travelers, and the Geiranger Road (1882–88) would eventually serve to bring the king and his entourage from Geiranger and up a thousand meters to the highest lookout point. There, the road met with the Old Strynefjell Road (finished in 1894). The Trollstigen Road (1928–38) and the Sognefjell Road (1936–38) were similarly built to encourage a flow of tourists. All of these are steep hairpin-bend roads in midwestern Norway. They were all preserved as late as 2009, and only after the TRP had started restoring them.

The explicit aim of the Road Administration is to encourage tourists to travel by car up through Norway and visit marginalized settlements, thus supporting them and establishing healthy tourism economies: "The aim is to strengthen commercial interests and settlements, particularly in rural regions."[8] It is further an explicit aim to display contemporary Norwegian architecture.[9] And last: "The traveler will experience Norway the way it is exhibited in tourist brochures and on postcards."[10]

The TRP represents the first time Norway has displayed an infrastructural project, a fact that signals a significant shift in the country's tradition for installing in nature. According to Raymond Tabata, already in 1989 tourism was only "second to oil as an item in world trade,"[11] making it thus somewhat stirring that a country flourishing on its own, as well as Canadian and U.S. oil reserves, now finds it timely to invest 3.4 billion NKr (US$560 million) on an extensive tourism infrastructure. Eighteen routes are in the making, six of which have been completed. Along the routes, the TRP is building a menagerie of amenities, from viewing platforms, rest areas, and lay-bys to paths, ferry quays, and visitor centers. About fifty-five of these are completed and thirteen are being built between 2011 and 2013, four of which are art installations. Counting all smaller and larger facilities, including the finishing of exiting Road Administration amenities (simple rest areas), 150 have been completed and about a hundred remain to be installed.

The project draws on already established tourist routes, expressly inspired by the American Blue Ridge Parkway in North Carolina and Virginia. And it is being marketed at a near global scale: The exhibit *Detour* has visited fifteen major U.S. and European cities, including New York and Washington, D.C. *Detour ASIA* has visited Shanghai, Beijing, and is opening in Guangzhou March 1, 2012, with a planned schedule for Japan and South Korea in 2012–13. The only publicized documentation of the project so far is the catalogue *Detour* published in four different editions since 2006.[12]

The most potent part of the TRP is the Geiranger–Trollstigen Tourist Route, comprising two of the late nineteenth-century mountain passes Geiranger and Trollstigen. It runs through the World Heritage Site West Norwegian Fjords—Geirangerfjord and Nærøyfjord in the midwestern part of Norway, resulting in a overexposure of Geiranger, a small settlement at the head of the Geiranger Fjord, which is chock full of tourist, cruiseships, cars, and hikers during the summer months. Geiranger appears as an open-air museum in these months: traditional Norwegian storehouses on pillars, wooden well houses, goats grazing, dry walls supporting the narrow road barely supporting the coaches full of tourists—all is accessible literally at an arm's length through the car window on the narrow one-lane road. The TRP does tap into the UNESCO area as it moves through Geiranger. The heritage museum in Geiranger displays the old farming culture of the region.[13] The TRP tourist, however, moving through in a car, will not necessarily be aware of this as she will be busy navigating the road and the logicstics of her travel.[14]

The West Norwegian Fjords—Geirangerfjord and Nærøyfjord were added to the UNESCO heritage list in 2005, on the grounds that the two fjords are "classic, superbly developed fjords" and distinguishable from other fjords already on the list (those in Japan and New Zealand). "The property displays a full range of the inner segments of two of the world's longest and deepest fjords."[15] The nomination document further states: "The two fjords represent one of the cradles of scenic tourism in Europe and have been able to cope with great influxes of tourists without losing any of their qualities."[16] The UNESCO preservation of the area already harbors incentives to preserve it as one of prolific tourism, as well as the scenic qualities of the area.

The Geiranger–Trollstigen tourist route is deeply embedded in the World Heritage area, both because the two roads themselves are preserved and because they have been the most important cultural conditions for subsistence in the

2. Flydalsjuvet viewing platform, Geiranger-Trollstigen Tourist Route. 3RW Architects, Smedvig Landskap AS. Photography by Sixten Rahlff.

region, which is populated with small settlements. But as the UNESCO document indicates, the roads serve to sustain another trait, namely scenic tourism, implying that there is a tradition for *viewing* involved in the listing of the region.

The TRP manages two viewing projects in Geiranger, both by architects 3RW and landscape architect Arne Smedsvig. One viewing platform, Ørnesvingen (the eagle bend), is a concrete cantilevered sheet placed off of the curb of the road, providing a zooming view out toward the Geiranger Fjord as well as down on the settlement. The other one, Flydalsjuvet (the Flydal gorge), facilitates tourists' presence at the rim of a gorge but includes an architectural structure, a restroom, that picks up on, and beautifully highlights, old building traditions by placing glass building blocks underneath an old log house structure. Here, the architecture underscores the TRP project as a play on authenticity. The building is a blend of old and new and demonstrates how heritage "distort[s] authenticity of conservation as an accurate revelation of the past," as Noah Nasser puts it.[17] This in-between state of preserving and creating heritage applies to the viewing areas as well, and to most of the TRP installations: Many of the lookouts already existed as crude wooden platforms or just worked-up spots along gorges, precipices, and rivers. Given the amount of architectural infrastructure now installed at the sites (i.e., pathways, cor-ten stairways, concrete paths, in addition to platforms), visitors are likely to stay within the designed structure and not contribute to the deterioration of the terrain, which is often a risk with specialized tourism.[18]

Flydalsjuvet is one of those iconic tourist sites established by early tourists, particularly Kaiser Wilhelm II, as early as the 1870s.[19] The viewing area itself, situated below the restroom, accentuates a lookout that has been featured in postcards since around 1900, an example of which exists in the U.S. Library of Congress. In this way, the TRP is perpetuating a viewing practice that was established during the nineteenth century in Europe and Norway. At the Geirangerfjorden–Nærøyfjorden World Heritage Site, one can say that heritage is being produced, as a global tourism practice is receiving a new element of intangible heritage, namely a viewing practice that has a two-hundred-year-old origin but is only now receiving a designed expression.

Panoramic Views

Throughout the eighteen tourist routes, diverse installations bring visitors close to nature. A few encourage them to move into nature, for instance, by following a yellow handrail leading up a hill at Nappskaret, Lofoten, a path made of railway sleepers leading down to a beach, or a cor-ten steel staircase leading the visitors to seaside slabs at Hellåga.[20] Were one to remain exclusively on the viewing platforms, however, one would find that the picturesque ideals of framed views are in full operation, but again by tacit intuition and thus mostly as proof of how deeply embedded in the European tradition the elevated viewing point has become.[21]

The TRP viewing strategies perpetuate a tradition that has privileged the view. Hence they imply a modern relationship to nature as entertainment inherent both to the picturesque tradition and the panorama tradition.

Sohlbergplassen (the Sohlberg place) on Tourist Route Rondane, designed by architect Carl-Viggo Hølmebakk, is a beautifully fluctuating platform fit between pine legs and hovering above ground on thin stilts. It unfolds as a landscape as it invites tourists to closely encounter both the structure itself, the ground beneath it, and the trees around it. Still it opens toward another landscape, a landscape in the most literal sense, as described by Rosalind Krauss as "a reduplication of a picture that precedes it."[22] Krauss was speaking in general terms of landscapes reproducing preconceived notions of what landscape is, but her definition takes on a literal form at Sohlbergsplassen. Before experiencing the view, tourists are invited to look at an older painting of the same view, reproduced on the railing of the platform: Harald Sohlberg's *Winternight in Rondane* (1914).

As a neoromantic painter, Sohlberg was well aware of the national romantic painting tradition that featured dramatic

3. Atna lake and Rondane seen from Sohlbergplassen, National Tourist Route Rondane; Carl-Viggo Hølmebakk AS, architect. Photography by Jørn Hagen for the National Public Roads Administration.

valley scenery mostly in the western part of the country. The first representations of the Norwegian landscape include sketches, prospects, and paintings executed by early nineteenth-century landscape painters who set out to discover their country as part of the national romantic upsurge that followed Norway's relative political independence from Denmark in 1814. One of the most seminal of these early paintings is Johannes Flintoe's *View from Bjønnstigvarden* (1819), a painting that is deemed a quintessential western Norwegian valley view. The travelers are portrayed admiring the welcoming valley below them before completing the descent from the mountain plateau. Thereby the activity of looking is being stressed as one of leisure and everydayness, but nevertheless one of grandeur. The painting is emblematic of the way Norwegian wilderness was portrayed and mediated in the middle of the nineteenth century and also for the many valley roads and views included in the project.

The viewing platform at the Trollstigen plateau reproduces these valley views, posing wide panoramas while also framing the road that made them possible even in the late nineteenth century. Carl-Viggo Hølmebakk's telescopic viewing device at Oscarshaug along the Sognefjell Tourist Route bears a structural kinship to the Claude glass, a tinted mirror used to frame and reflect the views.[23] These works are not the result of conscious efforts from the architects to explicitly tap into a tradition for viewing. Yet this framing and staging has roots as far back as the eighteenth-century discovery of the horizon. As demonstrated by Stephan Oettermann by the development for the "see-fever"

4. Viewing platform on molo by Askvågen, National Tourist Route Atlanterhavsvegen. 3 RW Architects; Smedsvig, landscape architect. Photography by Vegar Moen for the National Public Roads Administration.

in the mid-eighteenth century, a desire for the horizon and all-encompassing views permeated succeeding viewing practices and triggered the invention of the panorama, but it was initially realized by Goethe's description of the peaks of the Alps in 1779.[24]

Elements of a viewing tradition are placed haphazardly throughout the project. The centerpiece of the traveling *Detour* exhibit was a looking cabinet created by August Fuhrmann, similar to *The Kaiser Panorama,* a typology that gained tremendous popularity throughout Europe in the 1880 and '90s. This inclusion represents, however, a rather haphazard reference to the European viewing culture of the late nineteenth century. It was the cosmorama[25] that introduced Norwegians to mountain views in 1825, when an exhibit staged in Bergen showed recently painted landscape prospects, including *A View of Nærøydalen from Gudvangen* and *A View of Måbødalen* by Carl Peter Lehmann.[26] Johannes Flintoe would later present gouaches and prospects of landscape views a number of times between 1833 and 1845 also employing the cosmorama. These are the kinds of views that are reproduced as sights in the TRP, albeit never with an explicit reference.

There has been no tradition for the "true" panorama in Norway, yet it may be apt to compare the totalizing perspective of some of the viewing platforms to this medium that "had to be so true to life that they could be confused with reality."[27]

5. The viewing platform Stegastein at National Tourist Routes, Aurlandsfjellet. Todd Saunders, Saunders & Wilhelmsen, architects. Photography by Vegar Moen for the National Public Roads Administration.

The panorama sought to represent 360-degree scenarios seamlessly, even importing fake terrains, imitation terrain, between the platform and the panoramic screens. The produced feeling of being "in the middle of the landscape" that Bernhard Comment underscores is hypostasized in the real landscapes of the TRP.[28] These are landscapes that we actually *can* access. Nevertheless, we are still propped up on a *platform* in the middle of the landscape. The distance to nature is maintained in the viewing platform as media. They mark the place that one looks *from* as well as the view one is supposed to look *at*. Several of the platforms amplify their own character of being exactly platforms for viewing, most noticeably Bergsbotn by CODE Architects, which in addition to being cantilevered also features a platform within the platform as part of its construction principle, thereby hypostasizing its working principle of elevating the viewer.

The paradox is obvious for a touristic infrastructure that aims at bringing visitors into close encounters with nature. Dean MacCannell claimed that the modern tourist seeks relief from an existential alienation and thus seeks the authentic in other cultures.[29] His concept of "staged authenticity" presupposes that other cultures and places are assumed to provide an authenticity that one lacks in one's alienated modern life. One could easily suggest that nature and cultural heritage is the object of desire for the tourist, and thus discuss whether the TRP

6. Viewing platform at Bergsbotn, National Tourist Route Senja. Code architects AS. Photography by Jiri Havran for the National Public Roads Administration.

is providing any measure of authenticity. In "Authenticity and Commoditization in Tourism," Erik Cohen discusses whether commercialized tourism, then, risks losing authenticity, as the tourist becomes trapped in an encircling tourist-space.[30] The TRP is not strictly commercial, even though it subjects Norwegian views to heavy staging and marketing strategies.

Given its productive dynamics, its indisputable will to display, install, and choreograph Norwegian landscapes for view, while at the same time somehow leisurely insisting that it is only accommodating viewers, the TRP may easily gain authenticity on its own terms, corresponding to Cohen's pragmatic handling of the term: "In principle it is possible for any new-fangled gimmick, which at one point appeared to be nothing but a staged 'tourist trap,' to become over time, and under appropriate conditions, widely recognized as an 'authentic' manifestation of local culture."[31]

Building with Nature

In his foreword to the last edition of *Detour,* Karl Otto Ellefsen claims that the architecture of the TRP is firmly embedded in a Norwegian tradition of building with nature and relies specifically on the work of Sverre Fehn, the most acclaimed Norwegian modernist architect. Fehn's method rested in geometrically adapting the architectural structure to the dimension and existence of the landscape that harbors it, specifically trees and rocks. As the architectural structures of the TRP emerge, one could say that some refer to and thereby reproduce this legacy of building with nature. This tradition is however unknown to most visitors. What they relate to is thus hardly "the felt presence of the 'genius loci' throughout"[32] but rather the reconstruction and enactment of a tradition that is itself invisible to them.

It is furthermore not a simple truth that the TRP projects may be frictionlessly included in a Norwegian modernist tradition. It is certainly not true of Stegastein, an overscaled cantilevered platform running thirty-three meters into thin air, except for the fact that it is built in wood.[33] Similarly Manthey Kula's magnificent cor-ten steel structure in Lofoten bears little witness of the unassuming tradition of building into nature, except for its scale and the fact that the architects' intent is to shelter the visitor from views.[34] Still, in this context one can say that the TRP represents a reinvention of Norwegian scenery and of a tradition of building with nature that is known to consider nature as form and presence. This again can be established as a case of "emergent authenticity," as defined by Erik Cohen: "'Emergent authenticity' stresses one aspect or refers to one manifestation, of the wider phenomenon of 'invention of tradition.'"[35] The TRP *enacts* a tangible heritage of architectural practice by so consistently straddling nature with architectural form, even though the model for this enactment will not be known to the visitor.

TRP Programs and Competition

The Tourist Route Administration holds competitions for all major installations, some of which are limited to invited participants. Young Norwegian architectural offices are prioritized over landscape architectural offices, and the highly competent jury of five includes one artist, one architect, one landscape architect, the rector of the Oslo School of Architecture and Design, as well as a representative of the National Public Roads Administration, who is also a landscape architect. The Tourist Route Project includes an art line of fifteen major artworks, all managed by artist Svein Rønning. Intriguingly, the art line served as a corrective to the TRP at large. While the latter is intent on displaying "the emblematic image of Norway," the former is intent on deconstructing what Rønning refers to as the case of "managing Norwegianness" in the TRP at large.[36] Through the art line, he wants to force the TRP out of its own iconology.

The programs for the different projects give no constraints or guidelines as to how the projects should be conceptualized or designed. Yet a design culture has developed in the project, where viewing platforms tend to hypostatize and thus implicitly maintain viewing situations that are inherent to the European and Norwegian panoramic tradition. Also, a global material aesthetics prevails, as most projects employ glass, cor-ten steel, concrete, and wood, in addition to traditional Norwegian granite dry wall techniques.

The programs may include information on cultural and mythological heritage at the sites, thus encouraging designers to consider and include the cultural heritage of the region they are designing for. As an example, the program for a rest area at Børra at Andøya[37] thoroughly informs of the complex mythology of the area: the ghost of the sea (Draugen), who is said to go there to hold sermons; the Sami place of sacrifice Bukkekjerka (the buck-stone), already preserved as cultural heritage; and the particular geologic circumstances, which includes exposures of some of the oldest known rock on earth (primitive granite, more than 600 million years old). There is a lighthouse, a sinkhole, and Iron Age settlements in close vicinity to the rest area. Designers are encouraged to consider and include these aspects of the site, while required to establish paths to the lighthouse and the Bukkekjerka, which is sometimes used for weddings. In other programs, there may be no requirements of consideration to traits of the cultural landscape or heritage.

Vøringsfossen Viewing Area

An iconic point in the TRP is the Vøringsfossen, a waterfall that drops 182 meters off the edge of the Hardangervidde (the Hardanger plateau). The area has been a tourist site since the late nineteenth century, and it is worn down and dangerous; photographers have been known to fall off the cliffs while seeking optimal dramatic angles. An invited competition was issued in 2009 to encourage the preservation of the terrain and to facilitate safe access for tourists. Architect Carl-Viggo Hølmebakk's winning entry features a slender glass-walled visitors' center, a two-hundred-meter secured footpath along the rim, a pedestrian bridge connecting the plateau Fossatromma to the Fossli Hotel, and not least, several viewing platforms, one of which practically hovers above the floor of the gorge and the river. The project not only preserves the already existing and worn terrain of the rim, allowing the visitors to stroll amid its structures and to enjoy shifting views in motion, it also puts the viewer in "impossible" positions, allowing for bird's-eye views of the gorge not possible without mechanical aid. This amplification of the visitors' exposed position at the rim of gorges and abysses seems to be a global trend in tourism. Several of the TRP viewing platforms echo the Grand Canyon Skywalk at the West Rim. Albeit the Skywalk displays a more gaudy aesthetic, it allows visitors to explore the position of entering the expanse of air above the valley floor, beyond the border of the rim and wall. The situation of being exposed in terms of heights and chasms is amplified by providing a setting that is not possible without architectural facilitation.[38] The Vøringsfoss project, as well as several other TRP viewing platforms, restage the

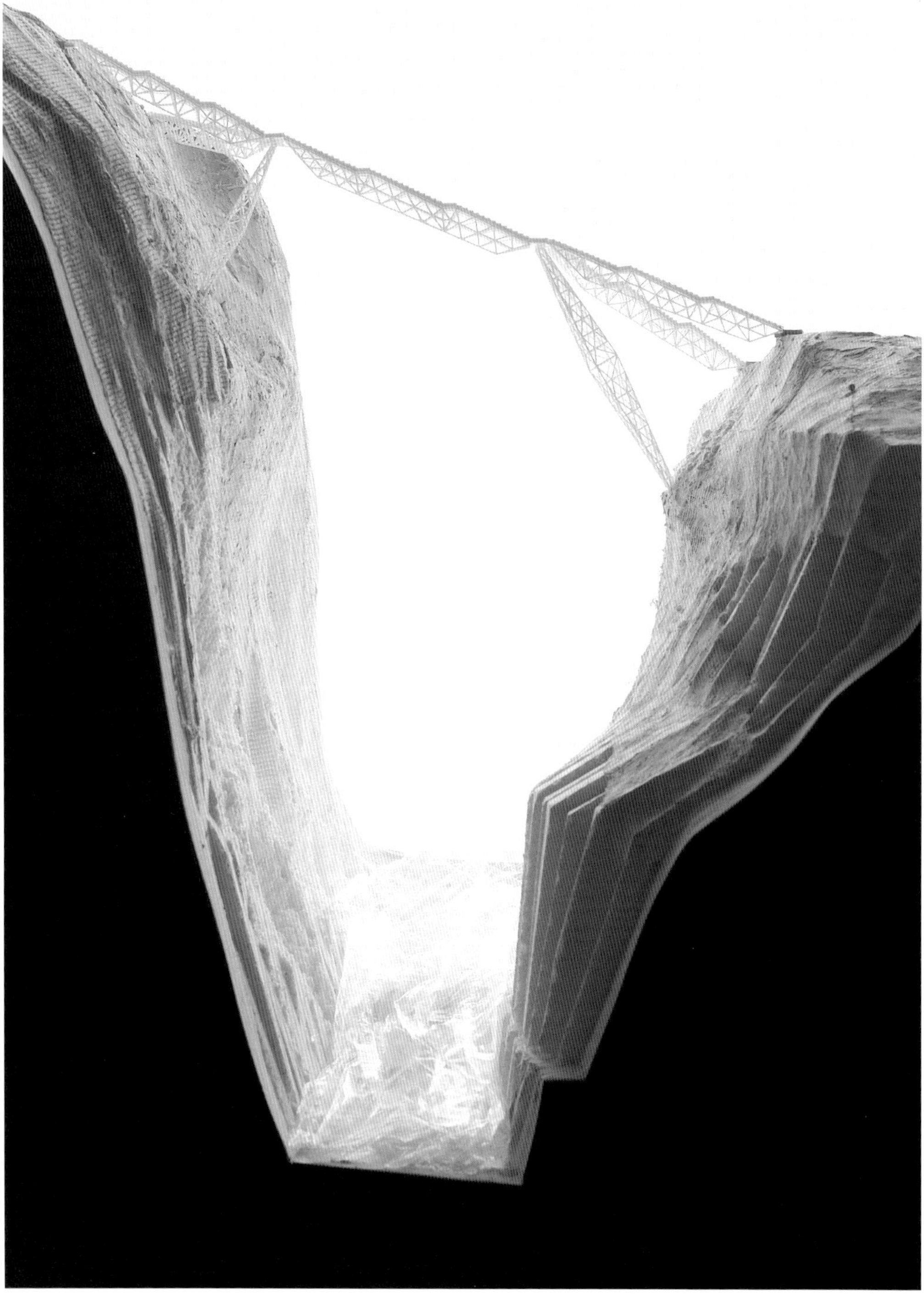

7. Drawing for competition entry. Carl-Viggo Hølmebakk AS / Andreas Holmer.

intimate relation to the drop or the precipice that was so integral to Burke's theory of the sublime. Thus they respond to the sensation-seeking tourist looking to be entertained by nature.

Carl-Viggo Hølmebakk's renderings for the competition echo earlier descriptions of the gorge as dark, frightening, and godforsaken.[39] He does, however, consciously connect to a long tradition of superstition and folktales integral to Norwegian nature conception and not the least to the post-1814 nation-building process.[40] Asbjørnsen and Moe's collection of fairytales greatly helped define the essences of "the Norwegian" in the decades following 1814. One of the competition visuals of the bridge alludes to the Norwegian folktale, part of the Asbjørnsen and Moe collection, of the three rams who defeat the troll while crossing a bridge on their way to summer pastures.[41] The connection to the folktale is obvious in the competition entry, but it mostly contributes to giving the jury a contextualized view of the concept. Some of these drawings have also been featured in news articles. Thus the perpetuation of nineteenth-century mythology pertains mostly to national evaluations of the routes and are not yet expressively available to the tourist.

However, when Hølmebakk chooses to include an enclosure for grazing goats, this is not a response to the program. The surrounding mountain areas have been part of the cultural landscape of centuries, and grazing goats, goats blocking the road, and goat cheese for sale are integral parts of the scenery. Hølmebakk's proposed staging of goats grazing has a precursor in the laying out of tourist routes in both Germany and the United States. As Thomas Zeller shows, the roadside landscapes of the Blueridge Parkway were heavily administered; people moved out of the areas, and pastures were constructed.[42] Thus again the TRP displays a consistent ability to unwittingly tap into traditions and trends.

Preserving Views

Norwegian legislation offers no protection of viewsheds equivalent to the U.S. State Coastal Management Program, which allows for designation and protection of Scenic Areas of Statewide Significance (SASS) in the United States. Viewshed protection is often connected to artistic expressions and traditions, as is the case with Olana, the Persian-style home and estate of Hudson River School artist Fredric Edwin Church perched atop a high hill near Hudson and Greenport, and prized for its architectural significance and scenic views.[43] Yet there is no specific view at or around Olana that can be rediscovered as painted landscapes in the Hudson River School of painting. The practice of painters, including Asher Durand as well as Thomas Cole and Fredric Church, was "never about representing what was, but

8. Model of bridge between Fossatromma and Fossli hotel. Carl-Viggo Hølmebakk AS, architect.

what the artist chose to see *as artist:* moralized pastiches of the past in the face of a troubling present."[44]

The SASS justification for preservation could apply to some of the views displayed by the TRP. The project favors landscapes that have already been mediated as part of a national tradition of painting, the most explicit one being Sohlberg's *Winternight in Rondane.* As is the case with Olana, however, the TRP does not necessarily reproduce exact views, but a mode of *attention* toward mountain scenery that has a significant precedence in Norwegian travel history. The TRP reproduces and maintains views that have been admired by tourists visiting Norway for centuries. Often these views have been made iconic by the practices of tourists themselves. Thus one could argue that the TRP is *performing* heritage, which is, as preservation theorist Laurajane Smith expands the term, "a process in which place, meaning, and identity [are] actively created and recreated by both the visited and the visitor."[45] In line with her argument, the entire TRP can be thought of as a "multilayered heritage performance" where "visiting, managing, interpretation, or conservation" are part of the practice.[46] What is mainly being "managed" and "conserved," however, is a particular viewing practice, not the landscape.

What makes viewing platforms like Stegastein by the Aurland Road and Askvågen along the Helgeland Road significant is really the way in which they enable visitors to connect with, experience, and be a living part of this Norwegian visual tradition.[47] Both sites mark the subject as a viewer of a carefully framed traditional viewing situation. However, the site for observation emerges as a place for a strong subject mastering its views, and thus the overwhelmed subject of the sublime is expelled.[48]

Passing Through

The TRP as a heritage performance, within or outside of the UNESCO heritage area, seems to be applicable to the notion of "passing through" noted by Ian Hodder and developed primarily in relation to the nine-thousand-year-old Turkish site Çatal Hüyük.[49] This is not a Western site, but the metaphor of "passing through" heritage seems to quite precisely describe the dynamics of the TRP as an infrastructural project.

Geiranger receives 160 cruise ships in a normal season (May through September), and tourism is the main means of subsistence for the 250 permanent residents. Ian Hodder's notion of passing through and Smith's interpretation of it as one that "reduces both locals and global visitors alike to the status of 'tourists,'"[50] thus seems particularly apt in the TRP, as a heritage performance enabled by movement and as an applied logic of "passing by." This reduction, Smith claims, is a "common consequence of the AHD [authorized heritage discourse] and the degrees to which it relegates visitors to the status of passive audiences," and as a consequence the narratives and meanings constructed by the performance of travel end up being so fluid that they "lack any solidity or consequence for the traveler."[51]

In the light of the particular viewing tradition conjured up by the TRP, one could argue that the visitors, to the contrary, are being *enabled* both as strong viewers and as drivers. These visitors enjoy a spectacle of both nature and architecture, and their production of identity and narrative takes place in a well-founded appreciation of contemporary design. They are adding to their design expertise by being "audience to the management and interpretive performances of the heritage site/museum management and interpretive staff."[52]

Global Art

As the TRP is evolving, its fifteen art projects are being added, although only one is as of yet installed: Steilneset in Vardø, where Louise Bourgeois and Peter Zumthor created a memorial "to the 91 persons convicted of sorcery and burnt at the stake in Finnmark County" during the seventeenth century.[53] More than any of the other planned art installations, *Steilneset* refers to local history, and one could argue that it is an extreme case of the production of cultural heritage, as it brings to light a repressed part of Norwegian history. It is, however, referred to as "a double offense" by Svein Harald Holmen, a local idealist working to prevent further depopulation.[54] The installation is conceptualized to make up for the witch-burning processes, but the locals do not at all relate to its formal expression. Also, neither local building material nor labor has been employed in this 80 million (NOK) project. Thus the TRP's expressed aim of boosting local economies has been counteracted. Nasser's ideal connection between "the preservation of the past" and tourist sites' ability to be a "resource for the modern community as a commercial activity" may have succeeded at the national level of tourism management, but not for the locals.

The dynamic between the individualized gaze of the tourist and the composition of a larger all-encompassing picture by the planning authorities might be relieved in these artworks. Only eleven of them have so far been commissioned. A handful of artists and land-art artists have been invited to take part in a later stage of the project. Among these are Mark Dion, Fischli & Weiss, Lars Ramberg, Marianne Heier, and Tue Greenfort.[55] The presence of Fischli & Weiss draws on their Swiss nationality. Their artwork at Tourist Route Valdresflye, to be installed fall 2012, will be an eighty-ton natural boulder placed on top of another, a five-meter-high sculpture. The sculpture is "formed" to a very small degree. By placing two found boulders on top of each other, they perform a minimal intervention at the site, and still the result is striking. Because Fischli and Weiss are Swiss and the sculpture built with elements from the very bedrock so important to Norwegian identity, an interesting connection is established by an artist from another nation whose wilderness was made passable in the name of national identity. A twin project will most likely be placed in Hyde Park, which will complete the triangular structure of Norway, Switerzerland, and England in the name of tourism 150 years ago. English bourgeois travelers were seminal to the discovery of both the Norwegian and Swiss mountains, and when Norway started developing its mountain roads, it was in explicit competition with Switzerland for tourists.[56]

Norwegian nature, the art line claims, is not about splendid views any longer; it is about resource management, about water, oil, gas, and tourism as a means of sustenance. The national conception of natural resources and scenery needs to be looked at as part of global resource management, a biased insistence on decentralization within a project that so fundamentally performs a reinforcement of exactly the specific Norwegian nature and local history. Typical of this perspective, Mark Dion's sculpture, soon to be installed in a hillside between Aurdal and Lærdal in Western Norway,[57] is called *The Den* and features a bear sleeping atop a garbage pile of used objects from the Viking Age until today.

The project most explicitly referencing a global context is Marianne Heier's projected work along the Atlantic Road. This road features a series of magnificent bridges leaping and throwing themselves between the islands.[58] Built 1983–89 and 8.2 km long, the road connects communities on both sides of the small archipelago that it spans with eight bridges, supported by several landfills.

The TRP competition wanted to focus on the road as communication and the bridges as communication links. The artist responded by suggesting an eighty-meter-deep well in the south end of the Atlantic Road, as well as nine identical wells in the Malawi Desert. These will be classic wells with traditional hand pumps. In cooperation with UNICEF, the wells would be built adjacent to villages with poor water supplies. It is of great importance both to the artist and to the TRP that, for the African users, the water pumps not be publicized as anything other than generic pumps already used by UNICEF.[59] Marianne Heier stresses the fact that the water pump installation connects to the flow of water as one of the shared resources, and one that could be better shared globally. Also, she wants to call attention to the flow of tourists encouraged by the TRP scenic routes. Leisurely, visually informed scenic travel is a rare kind of travel in a world where billions migrate for more pressing existential reasons. Finally, she also wants to strike a blow to the idea of the national by drilling the earth's surface for water. Thus she shares UNESCO's concern for the "globality" of the territorial geography and geology in the Geiranger–Nærøyfjorden World Heritage Site.

Erik Cohen poses "tourist art" as a "salient example of the commoditization of a range of cultural products through tourism."[60] The TRP art, however, has not been part of a local tradition. Also, it is not produced by local artists but by highly internationalized artists. Further, it is produced for a specific purpose and seeks to comment on the situation of which it is part. The TRP art is, furthermore, subject to another trend, that of "thinning"—a concept put forth by Rem Koolhaas in an attempt to conceptualize the fact that an increased appropriation of territories through infrastructure, industrial facilities, art, and preservation happens without a corresponding condensation of human presence.[61]

Invented by the elite, by kings and English noblemen, the Norwegian *terroir* is still on display for the global elite, as it is only a tourist of means that can afford driving through Norway. The TRP project seems to escape established categories for organized tourism. It reveals a quite random blend of practices, where considerations of local heritage are mostly left to designers' choice and intuition. Still it unwittingly taps into historically defined social practices and a globalized design language. It unfolds as a theatrical staging of the Norwegian landscape that brings the viewing practice from the absorbed state of the sublime to a more controlled position as viewer. This infrastructural viewing machine gains status as "medium" or "storyteller" in Silverstone's vocabulary. It addresses a somewhat "specialized" viewer (Cohen), the one in favor of cutting-edge architecture and northern "wilderness." Still this viewer is becoming aware of the staging of the nature she thought to be pristine and untouched, and gains awareness of her own practice of viewing. Thus the project has an aspect of what *smudgestudio* would call "civic pedagogy": "You could say that developing an attention to one's own imagining and production of meaning supports, and might even be analogous to, developing attention to one's own sense of and sensations of learning."[62]

Biography
Janike Kampevold Larsen holds a Dr. Art. in literature and is associate professor at the Institute of Urbanism and Landscape at the Oslo School of Architecture and Design. She has done research and written on the Norwegian Tourist Routes as part of the research project *Routes, Roads, and Landscapes: Aesthetic Practices en route, 1750–2015* . She is currently working on the Norwegian North and establishing a teaching program in Tromsø, Norway, that will address the emerging challenges in the circumpolar territories and landscapes. She recently coedited *Routes, Roads, and Landscape* (2012) and is working on a book on the Tourist Route Project.

Notes

[1] Michel Houellebecq, *The Map and the Territory,* trans Gavin Bowd (London: William Heinemann, 2011), 291.

[2] Ibid., 284.

[3] Tourism in Norway started in the eighteenth century, and many travellers published their observations.

[4] Quoted by Noah Nasser in "Planning for Urban Heritage Places: Reconciling Conservation, Tourism, and Sustainable Development" *Journal of Planning Literature* 17 (May 2003): 471.

[5] The term "authenticity" was developed by Dean MacCannell in *The Tourist: A New Theory of the Leisure Class* (Berkeley and Los Angeles: University of California Press, 1999) and has been discussed and problematized in a series of works, not the least Erik Cohen's "Authenticity and Commoditization," *Annals of Tourism Research* 15 (1988): 381–86.

[6] Susan S. Fainstein and David Gladstone, "Tourism and Urban Transformation: Interpretations of the Urban Tourism," in *Cities in Transformation—Transformation in Cities: Social and Symbolic Change of Urban Space,* ed. O. Källtorp, I. Elander, O. Ericsson, and M. Franzén (Aldershot, U.K.: Avebury Ashgate Publishing, 1997), 120.

[7] Robert Silverstone, "Heritage as Media: Some Implications for Research," in *Heritage Interpretation, Volume 2: The Visitor Experience,* ed. David Uzzell (London: Belhaven Press, 1989), 141.

[8] See www.turistvegprosjektet.com.

[9] Foreword to *Detour: Architecture and Design along 18 National Tourist Routes in Norway,* ed. Nina Berre, 14–21 (Oslo: *National Public Roads Administration*/Norsk Form, 2006).

[10] This statement was to be found on a TRP webpage that no longer exists.

[11] Raymond Tabata, "Implication of Special Interest Tourism for Interpretation and Resource Conservation," in *Heritage Interpretation, Volume 28: The Visitor Experience,* ed. David Uzzell, 68.

[12] Nina Berre, ed., *Detour: Architecture and Design along 18 National Tourist Routes in Norway* (Oslo: National Public Roads Administration/Norsk Form, 2006).

[13] Examples are Geiranger Fjord Centre, which "displays and describes daily life in the small settlement from the past to the present. The multimedia show 'From mountain to fjord' shows the world site areas as it changes though the four seasons," http://www.geiranger.no/kalender/venueevents/5-norsk-fjordsenter. On view in this museum is also a plaster model of the Geiranger Road, which received a gold medal on the World Fair in Paris 1990.

[14] Nasser also refers to concerns that local residents are becoming "spectacle" to the visitors' gaze. Nasser, "Planning for Urban Heritage Places," 473. This implication is avoided in Geiranger where nearly 100 percent of the population lives off of tourism and who close down and move to southern Europe during the winter.

[15] See http://whc.unesco.org/en/list/1195.

[16] "The West Norwegian Fjords, Norwegian Nomination 2004, UNESCO World Heritage List," 11.

[17] Nasser, "Planning for Urban Heritage Places," 470.

[18] Ibid., xx.

[19] See www.turistveg.no; search words "Ørnesvingen" and "Flydalsjuvet." The project, along with their nonwinning entry for Vøringsfossen, can also be viewed at 3RW's homepage: www.3rw.no.

[20] See www.turistveg.no; choose Tourist Route "Helgelandskyst Nord," then "Architecture."

[21] Another recent example is Christophe Girot's memorial at Der Invalidenplatz, close to Hamburger Bahnhof in Berlin. The installation is a solid staircase, bringing the visitor to a high point overlooking the plaza and thus metaphorically into a position to survey the complexity of historical references at the site.

[22] Rosalind Krauss, "The Originality of the Avant-Garde," in *The Originality of the Avant-Garde and Other Modernist Myths,* 151–71 (Cambridge, Mass.: MIT Press, 1986), 163.

[23] See www.turistveg.no; choose tourist route "Sognefjellet," then "Architecture" and "Nedre Oscarshaug."

[24] Stephan Oettermann, *The Panorama: History of a Mass Medium,* trans. Deborah Lucas Schneider (New York: Zone Books, 1997), 11.

[25] Cosmorama were "landscape scenes . . . mounted to form a semicircle and enclosed in a case. To view them the visitor stepped up to a matching semicircle of magnifying lenses set in front of the case." Oettermann, *The Panorama,* 69.

[26] Nils Messel, "Discovering the Mountains," in *Oppdagelsen av fjellet / The Discovery of the Mountains* (bilingual edition) (Oslo: Nasjonalmuseet for kunst, arkitektur og design, 2008), 127.

[27] Bernard Comment, *The Painted Panorama* (New York: Reaktion Books, 1999), 7.
[28] Ibid., 78.
[29] Dean MacCannell, "Staged Authenticity: Arrangements of Social Space in Tourist Settings," *American Journal of Sociology* 79, no. 38: 589–603.
[30] Cohen, "Authenticity and Commoditization," 373.
[31] Ibid., 380.
[32] Christian Norberg-Schulz, *Genius Loci: Towards a Phenomenology of Architecture* (New York: Rizzoli, 1980), 78–81.
[33] See www.turistveg.no; choose tourist route "Aurlandsvegen," then "Stegastein."
[34] See www.turistveg.no; choose tourist route "Lofoten," then "Akkarvikodden."
[35] Cohen, "Authenticity and Commoditization," 10.
[36] Svein Rønning, in conversation; used with his approval.
[37] Andøya is located north of the Lofoten Islands and is the northernmost island in the Vesterålen archipelago.
[38] See my reading of Stegastein viewing platform in *Detour: Architecture and Design along 18 National Tourist Routes,* ed. Nina Berre, 22–27 (Oslo: Norsk Form and Norwegian Public Roads Administration, 2006). This is truly an "impossible" position, a location one could never enter if it weren't for the platform. Also it is one of those locations dreamed of by nature wanderers, to always reach higher up and further out for a better view.
[39] Christian Asbjørnsen describes the Måbø Valley in Christian Tønsberg's collection of woodcuts entitled *Norge fremstillet i tegninger* (Norway displayed in drawings) (Kristiania, 1848). It is described as godforsaken, awe-inspiring, and gloomy.
[40] The year 1814 marks Norways's independence from Denmark. The next few decades show a rapidly increasing interest in Norwegian nature. Geologic surveys started in 1773, and painters residing in Berlin, among them Johannes Flintoe and J. C. Dahl, started making hazardous expeditions in Norwegian valley and mountain areas to document the wilderness.
[41] Per Christian Asbjørnsen and Jørgen Moe, "De tre bukkene bruse"/ ["The Three Billy-Goats Gruff"], in *Norske Folkeeventyr* (Norwegian folktales) (Christiania, Norway: 1841–44).
[42] Thomas Zeller, "Staging the Driving Experience: Parkways in Germany and the United States," in *Routes, Roads, and Landscapes,* ed. Hvattum et al., 125–38 (London: Ashgate 2011).
[43] "Catskill-Olana Scenic Area of Statewide Significance," www.nyswaterfront.com/SASS/PDF/Catskill_Olana.pdf. Cited in Fox, *Conflict and Common Ground: Analyzing the Relationship between Contemporary American Environmental and Historic Preservation Advocates.* Master of Science in Historic Preservation, Graduate School of Architecture, Planning and Preservation, Columbia University, 2008. Emphasis original.
[44] Ibid., 101.
[45] Laurajane Smith, *Uses of Heritage* (New York: Routledge, 2006), 71.
[46] Ibid., 3.
[47] See www.turistveg.no; search words "Askvågen" and "Stegastein."
[48] For a more detailed analysis of the fixity of the viewing situation within the TRP, see my article "Curating Views: The Norwegian Tourist Route Project," in Hvattum et al., *Routes, Roads and Landscapes,* 170–90.
[49] Ian Hodder, "Sustainable Time Travel: Toward a Global Politics in the Past," in *The Politics of Archaeology and Identity in a Global Context,* ed., S. Kane (Boston, Mass.: Archaeological Institute of America, 2003), 142.
[50] Smith, *Uses of Heritage,* 72.
[51] Ibid., 72.
[52] Ibid., 70.
[53] See www.turistveg.no, choose the "Varanger" road, then "Architecture."
[54] Remark at the Territorial Practices conference in Tromsø, January 26, 2012. For more information on his work, see www.sveinharaldholmen.com.
[55] Very few of the art projects can be mentioned here, as most are not yet fully conceptualized.
[56] A founding member of the *Norwegian Tourist Association,* the main promoter of new roads in Norway at then end of the nineteenth century, F. W. Heyerdahl in 1881 promoted the building of Norwegian mountain roads to strengthen Norway's position as a tourist nation in relation to Switzerland: "It is not only to its magnificent nature Switzerland owes these many travelers, but certainly also to those accommodations with which one has equipped those places that travelers wish to go, in addition to the endeavors made to ease access to them." *Den Norske Turistforenings Årbok* (Kristiania, 1881), 167–68.
[57] See www.turistveg.no; choose "Aurlandsvegen."
[58] See www.turistveg.no; choose "Atlanterhavsvegen."

[59] The pumps are developed in Kenya and produced in India. They are mechanical and can be repaired onsite; parts are easy to get most places.
[60] Cohen, "Authenticity and Commoditization," 381.
[61] See http://www.scribd.com/doc/54426970/Preservation-and-Thinning-Rem-Koolhaas.
[62] See http://www.extrememediastudies.org/extreme_media/14_infrastructure/norway_pop.html.

1. Taco Cabana, September 2006. Image courtesy Sage Kitamorn.

Winnie Won Yin Wong

Ambience as Property

Experience, Design, and the Legal Expansion of "Trade Dress"

Who authors and owns a space? Who authors and owns its appearances and sensations? Who, in turn, has the right to preserve or transform it? Disputes over the authorship and ownership of architectural design in the realm of intellectual property law can indicate the limits of these questions and delineate one important battleground on which they are contested. In 1990, under the obligations of the Berne Convention, American copyright protection was extended to architectural works for the first time.[1] Under the act, a Yale University architecture student, Thomas Shine, sued David Childs, a partner at Skidmore, Owings & Merrill. Shine alleged that Childs had copied his student work in Childs-SOM-Liebeskind's 2003 unbuilt design of the New York Freedom Tower—work that Childs had indisputably seen when he served as a guest critic for Shine's review at the school in 1999.[2] The case was settled out of court in 2006, after the court dismissed one of the defendants' two motions for summary judgment, thereby allowing Shine's suit to go forward.[3] In writing his opinion, Judge Mukasey laid out the legal precedents for establishing originality in architectural design and argued that reasonable observers might well find substantial similarity between Shine's and Childs's designs.[4] The lawsuit marked a rare moment when the tensions of hierarchy, privilege, and propriety within the architectural profession and its training apparatus were publicly revealed, but anyone cheering for the underdog—in this case, a young student defending his individual creativity against a strong and storied firm—would likely find dissatisfying the court's reasoning that, for Shine's work to qualify as original, "the level of originality and creativity that must be shown is minimal, only an 'unmistakable dash of originality need be demonstrated, high standards of uniqueness in creativity are dispensed with.'"[5] The legal bar for owning a design as a copyright is, in other words, far lower than what the culture at large would grant as creative authorship.

These issues are part of long-standing debates over the judgment of architectural forms and their referential statuses, narratives of invention and influence, and the intractable fact that the privileges of authorship often extend only to individuals, despite the highly collaborative and referential nature of design work. Beyond these questions of creativity (and its inverse, copying), another domain of intellectual property law

Future Anterior
Volume IX, Number 1
Summer 2012

has moved toward a very different conceptualization of design. Trademark law, which deals in the differentiation of goods in the marketplace (as opposed to the right to reproduce artistic works), has contemporaneously advanced a whole other set of issues that exceed the concerns and questions addressed in designer- and preservation-centric histories.

The legal developments discussed in this essay represent an emerging set of legal claims that expand the purview of intellectual property law into domains of design consumption rather than design production or preservation. The consequences of these developments for architectural design and preservation practices, and their role in commodifying, conceptualizing, and legalizing design, remain new and contestable, and this article is concerned with expanding the historians' understanding of the questions that can arise from this remarkable shift toward consumer-, commodity- and market-based questions in contemporary American legal culture.

This essay introduces, from a design history perspective, the legal expansion of a little-known category of intellectual property law called "trade dress" and examines its particular attempt to expand the purview of intellectual property law into the intangible qualities of spaces as they are experienced by the consumer. Trade dress is most succinctly defined as the "total image and overall appearance of a product" and, in the three cases examined in this article, is remarkable for formulating a totalizing conception of the designed commodity that attempts to codify as immaterial a thing as the "ambience" of designed spaces.[6] Although the concept of trade dress was incipient in U.S. trademark law in the Lanham Act enacted in 1946,[7] it was not until the 1990s that extensive legal activity developed around it. Several prominent cases, including a number heard by the U.S. Supreme Court, worked to expand the legal purview of trademark issues to encompass architectural, interior, product, and packaging design. This expansive understanding of design, I argue, coincided with broader attention to "experience design" in American urban and commercial culture as well as to the preservation of "landmark" urban spaces.[8] In this paper, I examine three trade dress cases from 1986 to 2003, laying out how "ambience" emerged as a legally codifiable commodity, a development that draws into its wake the ambiguities of the design and commodification of experience.

When consumer experiences can be used to fortify the legal and commercial ownership of a space, how does it shift the conceptual boundaries of the public sphere and what ought to be preserved in it? When the "ambience" of a space is legally defined as a trademark, how does it alter the public calculus in defining, evaluating, and preserving well-known and recognized building appearances and atmospheres? The extension of trademark law to the built environment through the category of the trade dress is, in some respects, a declaration of the right to preserve a totalizing environment and its public experience long before it is construed as historically significant. As larger and larger parts of the built environment, from restaurant interiors to urban squares, become legally defined as intellectual property through consumer experience, we need to examine the many vectors by which (public) preservation relates to the realm of (private) ownership.

In 1978, a taco stand named Taco Cabana opened in San Antonio, Texas. Popular and cheap, by 1985 it had expanded into a five-restaurant chain. In that same year, another Mexican restaurant chain named Two Pesos opened in Houston, Texas. One year later, Taco Cabana sued Two Pesos, alleging that Two Pesos had infringed upon its trade dress. In the suit, the attorneys for Taco Cabana described its trade dress in the following way:

> a festive eating atmosphere having interior dining and patio areas decorated with artifacts, bright colors, paintings and murals. The patio includes interior and exterior areas with the interior patio capable of being sealed off from the outside patio by overhead garage doors. The stepped exterior of the building is a festive and vivid color scheme using top border paint and neon stripes. Bright awnings and umbrellas continue the theme.[9]

By claiming that it exclusively owned as intellectual property this overall appearance that identified its restaurants to its consumers, Taco Cabana effectively claimed that no other restaurant selling Mexican fast-food could use this set of architectural and decorative arrangements to create the festive "atmosphere" it claimed to be unique to its restaurants. A district court jury ruled in Taco Cabana's favor, finding that Taco Cabana possessed this inherently distinctive trade dress, and the judge doubled their awarded damages. Translating the ruling into everyday language, he also ordered Two Pesos to alter the appearance of its existing restaurants and to display in them a white sign with one inch black letters that must read: "Notice: Taco Cabana originated a restaurant concept which Two Pesos was found to have unfairly copied. A court order requires us to display this sign to inform our customers of this fact to eliminate the likelihood of confusion between our restaurant and those of Taco Cabana."[10] The decision was affirmed by the Court of Appeals of the Fifth Circuit.

Taco Cabana's victory, along with its broad and seemingly banal description of its claimed intellectual property, was unexpectedly upheld by the U.S. Supreme Court in 1992, which

2. Taco Cabana, April 2009. Photograph by Derek A. Mathis.

affirmed the ruling that Taco Cabana did possess an inherently distinctive and hence protectable trade dress. This victory had critical effects, none more important than the commercial viability of the enterprises involved. In 1993, Taco Cabana purchased and absorbed all the Two Pesos restaurants. By 2000, Taco Cabana was describing itself as "the original Mexican Patio-Cafe" and, owned under the same corporate umbrella as Burger King, had expanded to more than 140 restaurants across the United States.

The principles of trademark law, of which trade dress is a subset, rely on a modernist conceptual binary that disaggregates the ornamental aspects of a product from its utilitarian ones. In order to facilitate competition in the capitalist marketplace, trademark law enforces the principle that, while the "function" of products can and should be copied (to encourage competition), its "decoration" in the form of trademarks should not be counterfeited (to prevent confusion). In American practice, the validity of a trademark is often classified by a "spectrum of distinctiveness" that provides increasing levels of distinction from (1) generic, (2) descriptive, (3) suggestive, (4) arbitrary, to (5) fanciful.[11] Notably, this runs exactly in opposition to the high modernist principle of "form follows function." That is, the more formally unrelated a trademark or a trade dress is from the product's function, the more distinctive and hence legally protectable it would be as intellectual property. This countermodernist dynamic was remarked upon by the design historian Frederick Schwartz, who pointed out in his study of early twentieth-century German design that

> the law created a new class of representations radically separate from those of the actual, material functioning of the economy. Trademarks and other market signs were of course part of this economy, feeding energies into the cycle of commodity exchange; but the law stipulated that the signs that circulated on the market draw a new and totally unrelated picture of the world of commodities for the consumer.[12]

Unlike traditional trademarks, however, which are registered through words and a drawing (i.e., a logo), trade dress consists of multidimensional and multimedia features, the conceptual boundaries of which remain undefined. Recognizing that "human beings might use as a 'symbol' or 'device' almost anything at all that is capable of carrying meaning," the growth of trade dress as intellectual property recognizes that it is not only words or images that bear meaning in the marketplace.[13] As such, in the Taco Cabana case, the jury had been instructed by the judge to consider that the restaurant's trade dress could well consist of

> the total image of the business. Taco Cabana's trade dress may include the shape and general appearance of the exterior of the restaurant, the identifying sign, the interior kitchen floor plan, the decor, the menu, the equipment used to serve food, the servers' uniforms and other features reflecting on the total image of the restaurant.[14]

This breathtakingly broad definition of an intellectual property exceeds any given element or feature of its design, or the configuration of particular elements, and thus potentially can encompass design elements that might change across an unspecified amount of time. Indeed, by 2002 Taco Cabana restaurants had introduced a "new look" to its interior quite different from those described at the time of its lawsuit.[15] The stepped exterior and the "festive" color scheme had been maintained, but key architectural elements such as the garage doors and the hexagonal building plan had been altered. On its corporate website, Taco Cabana described these new features as part of "the Cabana style," a crucial component of what it called "the Cabana experience." Taco Cabana's appropriation of the terms "style" and "experience" are significant here for implicitly connecting the design of its restaurants to the intellectual property once determined to be so distinctive and protectable under the law. With the implicit backing of legal discourse, "style" here becomes an idea whose scope transcends any of the given elements or features of the objects encompassed under it, and extends across a broad swathe of things, claiming to hold true (or at least distinct and recognizable) as it is experienced across time.

Since it established a broad and important precedent, the 1992 Taco Cabana decision marked the beginning of intensive trade dress legal action in America, turning trade dress into a highly active area of intellectual property development. Definitions of trade dress remain highly contestable today and vary from court to court. Few simple principles based upon purely formal categories (like color or material) can be distilled from the evolving precedents and the tests that might be applied. But the Taco Cabana case is remarkable because it brought within legal purview the intangible atmosphere and ambience of a restaurant as a unique commodity and property—like the idea of historic character and integrity in notions of urban heritage.

When the case was heard in 1992, Justice Scalia had debated the uniqueness of restaurant ambience with the attorney for Two Pesos. Scalia argued, "In the case of a restaurant, the trade dress is the product . . . you're paying some money for the food, and you're paying some money for the ambiance. . . . I mean Macy's might have a distinctive Macy's look about the store. But I'm not buying the store. I don't pay money to go and look at Macy's. I do pay money to go and sit in that restaurant in that particular environment, right?"[16]

When the attorney for Two Pesos countered that restaurant ambience was then more akin to packaging, Scalia retorted: "I don't think it is packaging. I think you're talking about the substance of what's being sold. You're selling atmosphere and food, the two of them. You can have wonderful food in a lousy atmosphere. I'm not going to pay as much money."[17] What Scalia is tapping into here is the inseparability of the commodity from the experience of consuming it, while relying on a presumption that the restaurant ambience that he finds such a distinct experience is shared by consumers at large. But how and under what circumstances ought this rise to the level of a protectable intellectual property, exclusively held? What of those consumers who might care for a department store's ambience as much as a restaurant's? Scalia's argument in *Two Pesos v. Taco Cabana* seemed to posit ambience-as-commodity in a way that gave the idea broad scope yet undefined parameters.

The Supreme Court was asked to revisit its Taco Cabana ruling in a 2000 case, *Wal-Mart v. Samara Bros.*, which dealt with designer appliqués on children's clothing, but which raised the issue of whether trade dress could be inherently distinctive (without requirement to prove that a "secondary meaning" had formed in the minds of consumers). In this context, and in light of many court interpretations of the Taco

Cabana case, the Taco Cabana ruling was raised again, and the Court returned to the question of whether the restaurant's ambience constituted product design and configuration or product packaging. Justice Ginsburg suggested there with considerable ambiguity as to whether a restaurant decor was product packaging or product configuration, while Justice O'Connor pointed out that it depended upon whether a dining experience was treated in the Taco Cabana case as part of the product or its packaging.[18]

Ultimately, writing for the unanimous ruling in the Wal-Mart case, Justice Scalia offered this summation:

> It seems to us that design, like color, is not inherently distinctive . . . we think consumer predisposition to equate the feature with the source does not exist. Consumers are aware of the reality that, almost invariably, even the most unusual product designs—such as a cocktail shaker shaped like a penguin—is not intended to identify the source, but to render the product itself more useful or more appealing.[19]

In other words, Justice Scalia is arguing here that a design can never be inherently distinctive as a means of identifying the commodity's origins. He raised instead the example of a "classic" glass Coca-Cola bottle: The bottle may constitute packaging for some, or a collectors' item for others, or something in between.

What, then, are the consequences of this legal parsing of the commodity for our understanding of design? Apart from a few collectors and design professionals, the American public, as Justice Scalia implies, has largely forgotten Emile A. Schuelke, the designer and patent holder of a 1930s silver-plated or gilt-silver penguin-shaped cocktail shaker for the Napier company.[20] At the same time the contour Coca-Cola bottle design, introduced in 1916 but redesigned and reissued many times over the past century, is widely considered a "classic." The question for a design historian to answer is whether this historical status is explained by Scalia's observation that product design largely fails to serve as a source identifier in the consumer's mind. Ought preservationists or historians defend the creativity of an Emile Schuelke or the shared consumer experiences—however intangible—of the changing but totally recognizable form of a Coca-Cola bottle? How, in other words, do we historically evaluate design under these dynamics of ownership?

Returning to the question of restaurant decor, Scalia wrote in the 2000 opinion that the case had "unquestionably" established the legal principle that trade dress can be inherently distinctive but that the decision was not going to be revisited because "the decor of a restaurant, seems to us not to constitute product design. . . . It was either product packaging . . . or else some tertium quid that is akin to product packaging."[21] Scalia's idiosyncratic use of this Latinism meaning "some third thing," fudges a distinction that had become unworkable.[22] It signals a breakdown in the function–ornament dichotomy and implies a need to redraw the existing parameters for understanding the designed commodity, while foregoing the opportunity to articulate an alternate set of distinctions. The adjustments in the Court's position from 1993 to 2000 over the nature of ambience as commodity point to the contradictions arising out of a wider understanding that commodification had grown to encompass not just products and things, or even spaces, but also the individual and idiosyncratic experiences of consumption construed as common ones.

The drive to formulate a legal understanding for the intangible ambience of restaurant in the Taco Cabana case is indicative of a larger trend, which the economists Joseph Pine and James Gilmore addressed in their well-known 1999 book, *The Experience Economy.* In it they described and advocated for the transformation of the American marketplace from a market of goods and services into one that offers "staged experiences," in which companies are "experience stagers," labor is "performance," and "the consumer is the product."[23] By this last pithy phrase, Pine and Gilmore were pointing out that the commodity of experience occurs within the consumer, an individual who has been offered "sensations" and who "has been engaged on an emotional, physical, intellectual, or even spiritual level" with memories that "linger on."[24] This conceptualization of the commodity as something tied to intangible qualities of sensation, experience, and memory is useful for understanding the emergence of trade dress as one way of evaluating the commodification of experience, providing a broader historical context for the contradictions revealed by the Supreme Court's recognition of ambience as a new and distinct kind of commodity.

Pine and Gilmore traced the origins of the experience economy to Walt Disney and the Walt Disney Company, crediting it for first "layering new levels of experiential effects onto cartoons" (e.g., by adding synchronized and stereophonic sound, color, audio-animatronics, etc.), and then creating the "living, immersive cartoon world" of Disneyland.[25] We might take the historical context a little further back, however, to a crucial precedent of Disney's theme parks: the World's Fairs, notable for institutionalizing thematically staged experiences tied up with the display but not purchase of commodities. Payment here, to echo Justice Scalia, was not for the displayed products but the structured experience of consuming them. In turn then, the well-known corporate pavilions of the 1939 New York's World's Fair (which served as a key precedent for Disney's EPCOT Center) might be retrospectively understood as sites in which companies were not selling their products but establishing their "trade dress."

Here it is interesting to consider another forgotten design, Raymond Loewy's Frozen Forest for the Chrysler Motors Building at the 1939 Fair. This was billed as a "frosted tropical experience" for the display of Plymouth, Dodge, De Soto, and Chrysler automobiles. Among the vehicles, tall artificial palm trees were placed throughout the display space, and cooled air blowing into the upper areas of the space would presumably cause the leaves of the palm trees to rustle. Artificial moonlight would also light up the frost that would presumably form from the atmospheric moisture that would condense on the trunks of the artificial trees. In a patent filing approved in 1941, Loewy described this invention as an "Apparatus for Decorating the Interior and Improving the Atmosphere of Enclosures," proposing its commercial use in restaurants, hotels, and ballrooms, to create "a feeling of comfort for the occupants of gathering places."[26] Though the design itself used existing air conditioning technology and essentially consisted of no more than artificial palm trees placed inside an air-conditioned space, Loewy claimed as his own innovation the "artistic arrangement of natural appearing creations in an enclosure, so as to simulate the appearance of a natural scene."[27] This is, if you like, an early intellectual property claim on the packaging of air as "ambience," extending well beyond particular design elements.

Such attempts to design and structure atmosphere and experience at the 1939 World's Fair served as an important precursor to the Disney Company's later commodification of atmosphere, nostalgia, and pleasure, and the much wider entertainment and licensing business concerns that are only incidentally connected to the company's far broader corporate identity. Taking Disney as the model of a firm's development of commodities from sensation to themed experience to the totalizing notion of a corporate enterprise producing a range of consumer experiences, notions like "style" or "spirit" would seem insufficient concepts to encompass its identity that "trade dress" might well replace. "Disneyfication," in this light, takes on a more complex dimension, of the expanding notion of commodity experience entailed within it, and the legal development toward consumer-centric concepts and ideas (intellectual properties).

More recently, Disneyfication has been widely applied to the transformation of New York's Times Square through the West 42nd Street Development Project, which centered around the codification of its diverse, visual, and layered ambience. This twenty-year project began in the 1980s with the criminalization of sex-related businesses and illegitimate theaters in the Times Square area through zoning, regulation, and police enforcement, and then transformed Times Square into a public theater in which the hybrid experiential commodities of "eatertainment," "infotainment," "advertorials," and "infomercials" were inaugurated.[28] This was a spectacular urban transformation in which the role of the Walt Disney Company was instrumental.

Issued in 1981, the 42nd Street urban design guidelines set out to create a Times Square ambience through formal attributes of urban and architectural design, each conceived as a critical component to an economy of experience, now brought to bear on the tourists' consumption of urban heritage. These guidelines were extremely detailed in scope and expressed a stated aim to preserve "Times Square's special ambience and visual character as a unique 'public room.'"[29] This included the articulation of a visual experience, including, "a vivid, around-the-clock pedestrian experience," an emphasis on "visual anchors at both ends of the block that through extravagant signage or brilliant architecture create dazzling gateways," and to provide visual experiences that were "compelling experiences in their own right."[30] To some architects, the guidelines were so detailed that they were said to virtually design the buildings.[31] In other words, design was situated as one element in a larger ensemble of legal-commercial practices intended to effect broad (social) concepts and ideas. It is significant that the "public" here explicitly entailed not only the urban citizen but also the consuming tourist from outside the city—for whom the urban heritage of Times Square would first be familiar through television, film, and news media. This intersection of the space of visual mediation with the public space of the city would further bring to the surface the emerging contradictions of corporate identity and its consumption.

The trade dress of one of the visual anchors of Times Square buildings became the basis of legal action in 2002, when Sherwood 48 Associates, the owner of the buildings at One and Two Times Square and 1600 Broadway, sued the Sony Corporation for allegedly infringing its trade dress in the 2002 film *Spider-Man*. As detailed in Sherwood's suit, in the trailer for the movie, the advertisement billboards that appeared on the buildings of Times Square had been altered using computer-generated imaging and electronic insertion technology. Specifically, the billboard for Samsung had been replaced in the film with a billboard for Cingular Wireless, a competitor of Samsung but also a major cross-marketing partner of the *Spider-Man* film. Sherwood also noted that the filmmakers had also altered the existing NBC logo with a USA Today logo.

In its suit, the building owners argued that, "This block of signs, which encompasses approximately thirty percent of all signage in Times Square, includes the majority of the most visible, most spectacular, and most expensive signs in the world."[32] The owners contended that Times Square was a "world famous, historic venue," forming "the backdrop for countless movies, television programs" and that its unique configuration and ornamentation were thus protectable trade dress. Since the buildings generate revenue almost entirely from their outdoor advertising,[33] they alleged that altered photographic, digital, filmic, and televisual images of their buildings would negatively affect the commercial value of that advertising space.[34] Hence, since the fictional *Spider-Man* movie advertising was an in-movie product placement that generated revenue for the production of the film, the filmmakers had theoretically sold "the same space" that the building owners owned. In other words, they contended that the buildings' trade dress—the "total image" of the buildings as a commodity—extended into the fictional space of a comic-book movie (and implicitly, to all its possible appearances in the world).

Sherwood's suit was dismissed by the Southern New York District judge. In a brief but surprisingly comical opinion, Judge Richard Owen starts out by narrating the situation at hand:

> Once upon a time, at a gathering of many thousands in New York's Times Square for the "World Unity Festival," the crowd was murderously attacked by the jet-powered Green Goblin, who was, however, eventually put to flight by the timely arrival of Spider-Man. . . . This event was memorialized on film, which swept the almost totally advertising-encrusted buildings surrounding Times Square, including New York City's Municipal Building (with an added balcony) which had been digitally removed uptown to be on the west side of the Square.[35]

By describing the fictional movie in the tenor of historical fact (even calling the film a "memorialization" and the filmmakers "memorializers"), Owen thus mocks the owners of the endlessly changing Times Square billboard buildings for claiming permanent and expansive rights of intellectual property. By adopting the narrative of the film in his own written opinion, Owen countered the building owners' totalizing claim upon the visual world of film fiction, reminding us in a grandiose summation that "what exists here is for artistic purposes." In other words, he suggests that the filmmakers held the artistic right to transform the appearance of a Times Square building (as an in-movie product placement) in order to fund the making of

3. Samsung billboard, Times Square, July 2008. Image courtesy Juriaan Persyn.

4. Times Square, December 2011. Image courtesy George Rex.

their film. Owen argued, therefore, that the filmmakers' movie was protected as free speech under the First Amendment.

In contrasting the intellectual and creative work of the filmmakers with the claimed intellectual property of building owners, Judge Owen's opinion reveals the competing historical dynamic of "property" applied to increasingly intangible commodity forms. To Sherwood's argument that the filmmakers had "trespassed" their property by electronically altering photographs of it, Owen dramatically scoffs: "trespass?—bouncing a laser beam off a building to create a digital photograph? Light beams bounce off plaintiffs' three buildings day and night in the city that never sleeps."[36] In other words, the *Spider-Man* suit and Judge Owens's opinion reveal two forces at work: the modernist privilege of artistic intention trumping ownership (public or otherwise), and a consumer-centered basis for the ownership of "appearances" that restricts its transformation. In the experience economy, what becomes increasingly in contention is not the ownership of copyrights but the very limits of images and appearances themselves.

The notion that fetishism totalizes the space of commodity exchange, and not just the commodity, is not new: Walter Benjamin's project on the Arcades might be seen as an attempt to catalogue their "ambience." The historian Jean-Christophe Agnew has linked the special theatrical value of Times Square to the fundamental economic conventions of the consumer's suspension of disbelief. Anticipating Scalia's thought that the consumer is often paying for more than just one thing, Agnew argues of the earlier Times Square's sexual and theater economy: "what is bought is both the fantasy and the right to the fantasy. . . . The customer accepts: first to use the price of his admission to conceal from himself the extent of the work and risk he must contribute; and second, to use the offered fantasy to conceal from himself the price of admission."[37]

As with Scalia's "tertium quid" conundrum, the difficulty here is in establishing the imaginary space in which consumption takes place. The legal codification of ambience as intellectual property reaches into that imaginary space in a remarkable way, placing measures of distinction and ownership upon the consumers' space of fantasy and his surrender of the payment of admission. While Pine and Gilmore's account of the advanced commodity of "staged experiences" may be no more new than the idea of the simple payment for admission, from Taco Cabana to Times Square, both this fantasy and its value have taken on a rather more solid form. At a Taco Cabana, the consumer paid for the food but, according to the Supreme Court, bought some other thing, a distinct ambience. At fairs and theme parks, the consumer paid admissions for a structured experience of corporate identity, immersing themselves in the product-worlds of companies without the purchase of those products. In post-Disney Times Square, the consumer does not pay at all but rather enters a "free public room," in which advertising light beams "trespass" into their consuming space. This is a trespass that some allege extends into fictional worlds as well, for where property is conceived of as appearances, so too can the appearances of property be owned and consumed.

The legal expansion of trade dress in recent decades thus brings to a crisis some long-standing historical dynamics. Where it is reasonable to see the notion of property extended to the mental labor of the artist and designer through the "intellectual" property of copyright and patents, so too would the purview of trademark law extend quite predictably into the mental space of the consumer. Where the boundaries of ownership are blurred by claims on common consuming experiences, so too are the property boundaries of physical buildings and urban spaces blurred with the fantasy light worlds of film, television, and multimedia. Agnew reminds us that the older, pre-Disney Times Square offered fantasy worlds for sale as well, and that the prices of admission consumers once paid for a dinner with ambience or for a flickering film had an excessive value, too. The "new" prices of admission are now conceived of as an experience of a company's trade dress or identity, allowing property owners to claim that by merely appearing a certain way they now hold a certain intellectual property. The legal bar for establishing a connection of trade identity to the consumers' experience and memory is thus as absurdly low as the originality required for the ownership of copyright. Hence, a Samsung product placement in the fictional "Times Square" of *Spider-Man* is defended as "free" speech, just as the New York advertising room of Times Square is made free and public. Here what is brought to crisis is not the mere commodification of everything but the making a fetish of values such as creativity, experience, and memory in order to defend that commodification.

Since the 1980s, American artists have deftly navigated the twentieth-century expansion of copyright law to works of art through the development of "appropriation" as a critical strategy, allowing them in some cases to challenge the boundaries of property in ways unanticipated by the law. Perhaps designers and preservationists, too, may soon need to find new means to address the legal expansion of trademark law into arenas of spatial experience, and to challenge some of the conceptual parameters of cultural heritage and ownership. As existing property is legally defined already by its total image and feel, so will the limits of its future transformation or preservation be altered. Working with this new space of commodity and fantasy will require new conceptions of authorship, design, and preservation.

Biography

Winnie Won Yin Wong is a Junior Fellow of the Harvard Society of Fellows, working in the history of modern and contemporary art with a focus on China. Her theoretical interests include transnational issues in authorship, creativity, and copying, and her research is concerned with questions at the intersection of art and intellectual property. She is at work on a book entitled *After the Copy: China and the Readymade,* a study on conceptual art and the world's largest production center for hand-painted art products.

Notes

An early version of this paper was presented at the Market of Effects symposium at the Yale School of Architecture in March 2007. I am grateful to respondents for their comments, to Arindam Dutta and Jordan Sand for their comments on earlier drafts, and to Peter Karol for our discussions on trademark law and trade dress.

[1] Architectural Works Copyright Protection Act 17 U.S.C. § 102–§ 1,332 (1990).

[2] United States District Court for the Southern District of New York, *Shine v. Childs and Skidmore, Owings, Merril,* 382 F. Supp. 2d 602 (2005).

[3] Jeffrey Brown, "Too Close For Comfort," *Architect* (November 2007): http://www.architectmagazine.com/architecture/too-close-for-comfort.aspx.

[4] Shine v. Childs and Skidmore, Owings, Merrill (2005) at 610, 615.

[5] Ibid., 610, quoting *Folio Impressions, Inc. v. Byer California,* 937 F. 2d 759, 764–65 (2d Cir.1991), quoting *Weissmann v. Freeman,* 868 F 2d. 1313, 1321 (2d Cir.1989).

[6] *Blue Bell Bio Medical v. Cin Bad, Inc.,* 864 F. 2d 1253, 1256; (CA5 1989). Also, it "involves the total image of a product and may include features such as size, shape, color or color combinations, texture, graphics, or even particular sales techniques." *John H. Harland Co. v. Clarke Checks, Inc.,* 711 F. 2d 966, 980 (CA11 1983).

[7] Lanham Act, Section 43(a), 60 Stat. 441, as amended 15 U.S.C. Section 1125 (a).

[8] Graeme B. Dinwoodie, "The Death of Ontology: A Teleological Approach to Trademark Law," *Iowa Law Review* 84, no. 611 (1999): 621–39. Mark A. Lemley, "The Modern Lanham Act and the Death of Common Sense," *Yale Law Journal* 108, no. 7 (1999): 1687–715. Thomas J. McCarthy, "Lanham Act Section 43 (a): The Sleeping Giant Is Now Wide Awake," *Law and Contemporary Problems* 59, no. 2 (1996): 45–74.

[9] *Taco Cabana International v. Two Pesos,* 932 F. 2d 1113, 1117, June 11, 1991 (CA5 1991).

[10] Tracey Taylor Woodard, "Judge Details Changes in Two Pesos concept–Trade Dress Misappropriation Suit Brought by Taco Cabana," *Nation's Restaurant News* , January 23, 1989: http://findarticles.com/p/articles/mi_m3190/is_n4_v23/ai_6996322/.

[11] The "spectrum of distinctiveness" was formulated by Judge Friendly in 1976 in *Abercrombie & Fitch Company v. Hunting World, Inc.,* 537 F.2d 4 (1976).

[12] Frederic J. Schwartz, "Commodity Signs: Peter Behrens, the AEG, and the Trademark," *Journal of Design History* 9, no. 3 (1996): 173–75.

[13] *Qualitex Co. v. Jacobson Products Co.,* 514 U.S. 159, at 162 (1995).

[14] *Two Pesos v. Taco Cabana,* 505 U.S. 763 1 App. 83–84.

[15] Taco Cabana public website "Cabana Style," copyright 2002, www.tacocabana.com.

[16] Oral transcripts, in the Supreme Court of the United States, *Two Pesos v. Taco Cabana,* April 21, 1992.

[17] Ibid.

[18] Oral transcript, in the Supreme Court of the United States, *Wal-Mart Stores, Inc. v. Samara Brothers, Inc.,* January 19, 2000.

[19] *Wal-Mart Stores, Inc. v. Samara Brothers, Inc.,* 529 U.S. 205 (2000) at 213.

[20] E. A. Schuelke, Cocktail Shaker or Similar Article, USPTO Reg. No. 101,559 (October 13, 1936). See also Joanne Dolan Ingersoll, ed., *Cocktail Culture: Ritual and Invention in American Fashion 1920–1980* , exh. cat. (Providence, R.I.: RISD Museum of Art, 2011).

[21] *Wal-Mart Stores, Inc. v. Samara Brothers, Inc.,* 529 U.S. 205, at 213 and 192, March 22, 2000.

[22] Lars S. Smith, "Trade Distinctiveness: Solving Scalia's Tertium Quid Trade Dress Conundrum," *Michigan State Law Review* 243 (2005): 243–308.

[23] Joseph Pine and James H. Gilmore, *The Experience Economy: Work Is Theatre and Every Business a Stage* (Boston: Harvard Business School Press, 1999), 11–12.

[24] Ibid., 12.
[25] Ibid., 3.
[26] United States Patent Office 2,251,706, Apparatus for Decorating the Interior and Improving the Atmosphere of Enclosures, Raymond Loewy, New York, N.Y., assignor to Chrysler Corporation, Detroit Michigan, application filed May 24, 1939 (Serial No. 276,477), patented Aug 5, 1941, CL. 62–140.
[27] Ibid., 1.
[28] The Times Square Business Improvement District regulates garbage, crime, and homelessness in an "autonomous security zone." Mark Sussman, "New York's Facelift," *The Drama Review* 42, no. 1 (Spring 1998): 2–4. See also Tom Burr, "Sleazy City: 42nd Street Structures and Some Qualities of Life," *October* 85 (Summer 1998): 93, quoting *Hells Kitchen Development News*, September 1997.
[29] Lynne B. Sagalyn, *Times Square Roulette: Remaking the City Icon* (Cambridge, Mass: MIT Press, 2001), 190.
[30] Sussman, "New York's Facelift," 36.
[31] Sagalyn, *Times Square Roulette*, 192, quoting James S. Russell, "Golddiggers of '84?" *Architectural Record* 177 (June 1989): 127.
[32] Costantini, Gulia and Morris, LLP, *Sherwood 48 Associates and Super Sign Company v. Sony Corporation et al.*, Civil Action no. 02 CV 2746, at 4 (2002).
[33] Louis Brill, "One Times Square, Signage in the Crossroads of the World," *Sign Industry*, May 30, 2002: http://www.signindustry.com/led/articles/2002-05-30-LB-TimeSquareOne.php3.
[34] Costantini and Morris, *Sherwood 48 Associates v. Sony*, at 22–23.
[35] Sherwood 48 Associates and Super Sign Company v. Sony Corporation of America et al., 02 Civ. 2746 (2002)
[36] In 2003, the 2nd Circuit Court of Appeals ruled that this balancing of the First Amendment and the Lanham Act ought to be reconsidered and remanded that aspect of the case to district court. *Sherwood 48 Associates et al. v. Sony Corporation of America et al.*, 76 Fed Appx 289 (2003).
[37] Jean-Christophe Agnew, "Times Square: Secularization and Sacralization," in *Inventing Times Square: Commerce and Culture at the Crossroads of the World*, ed., William Taylor (New York: Russel Sage, 1991), 8.

1. Fatīh Mosque, Istanbul, begun mid-fifteenth century by Sultan Mehmet II. Photograph by author.

Janna Israel

A History Built on Ruins

Venice and the Destruction of the Church of the Holy Apostles in Constantinople

When the Ottoman Turks took control of Constantinople from the Byzantine Empire in 1453, Sultan Mehmet II ordered the construction of a great new mosque and charitable precinct, or *külliye,* in Istanbul on the site of the former Church of the Holy Apostles.[1] Begun by Constantine in the fourth century, the church had been a nodal point for Byzantine ceremony, where relics of important Eastern Orthodox saints were housed along with the bodies of emperors credited with securing those relics.[2] The relics conferred spiritual and political legitimacy on the dynastic claims of the emperors newly rooted in Constantinople. When the Venetians sacked Constantinople during the crusade of 1204, many of the relics from the Church of the Holy Apostles were dispersed in the West. Since its establishment as a colony of the Byzantine Empire in the fifth century, Venice had a tenuous relationship with Constantinople as subject, enemy, and trading partner. As Venice prepared to mount a Crusade to Constantinople in 1464, this time against the Ottomans, the Republic mourned the destruction of the church. Following the destruction of the Church of the Holy Apostles, the image of the church became increasingly politicized.

This article explores the shaping of the biography of the Church of Holy Apostles as the Ottomans supplanted Byzantine hegemony in Constantinople and as Venice positioned itself as Byzantium's worthy heir rather than its rival. In his compendious *Realms of Memory,* Pierre Nora traced the ontology of places that became *topoi* for French identity through political and historical change. He defined a "place of memory" as any significant entity, whether material or nonmaterial in nature, that by dint of human will or the work of time has become a symbolic element of the memorial heritage of any community.[3] Drawing on Nora's work, I argue that the Church of the Holy Apostles was reproduced in Venice through traces of its relics and architectural forms as the Republic again redefined their relationship with the Byzantine Empire, claiming inheritance over its patrimony.

Byzantine descriptions of the Holy Apostles have enabled scholars to reconstruct the plan and elevation of the church. Writing around 1200, Nicholas Mesarites described how the builders "constructed circular wheels, extending to them four well-fitting curves called slings [pendentives]; and they [the wheels] likewise received five domes; but the architect

Future Anterior
Volume IX, Number 1
Summer 2012

2. Map of Constantinople, William R. Shepherd, *Historical Atlas* (New York, Henry Holt and Company, 1923).

arranged the middle dome in a reverent way, so it would project and reign over all."[4] Based on similar descriptions of the church's cruciform plan, five domes, and mosaic decoration dating back to the sixth century, two manuscript illuminations dated to the tenth and eleventh centuries, each depicting a building with four smaller domes arranged around a larger central dome, have been identified as representations of the Holy Apostles.[5]

The Church of the Holy Apostles in Constantinople had ostensibly represented the relationship between Byzantium and the Republic well before its destruction. From accounts of its cruciform plan, five domes, and mosaic decoration many scholars have argued that the Holy Apostles served as the principal prototype for the basilica of San Marco in Venice.[6] Built during the ninth century and rebuilt twice before the twelfth century, San Marco was founded to accommodate the relics of St. Mark the Evangelist after they were taken by the Venetians from Alexandria in the ninth century. As a newly formed Republic at the time of the ninth-century theft of St. Mark from Alexandria, the relics of one of the gospel's authors offered Venice credibility as a divinely ordained Republic, worthy of its prosperity. As the emperors guarded the relics of the apostles at the Church of the Holy Apostles, the elected figurehead leader of Venice, the doge, would oversee the relics of Mark the Evangelist, a saint considered to be an apostle, at his titular basilica.

Richard Krautheimer proposed that Greek architects, who appropriated the general plan and elevation of the Holy Apostles, designed the basilica of San Marco while later scholars emphasized the political affiliation embodied by the two structures.[7] According to Juergen Schultz, San Marco not only copied the footprint and the profile of the Church of the Holy Apostles, but after the Crusade of 1204, when objects spoliated from Constantinople, including columns, porphyry, and the four bronze horses were incorporated into the façade of the basilica in Venice, the Piazza San Marco reproduced the appearance of the Hippodrome of Constantinople.[8] The entire ducal complex in Venice would have pointed to Roman imperial and Christian authority at a time when Venice forcefully claimed independence from the rule of Byzantium. More recently, Marina Falla Castelfranchi specified that the third construction of the basilica of San Marco in the eleventh century drew on the Holy Apostles to express an allegiance to the Byzantine emperors at a time when Venice's neighbor to the north, Aquileia, tried to assert its power over the Italian peninsula.[9]

Centuries later, as the Eastern Empire receded, Sultan Mehmet's mosque also drew on forms associated with the Byzantine Empire. The mosque has undergone extensive renovations due to earthquake damage, but based on early views of Istanbul, a large central dome supported by subsidiary domes around the circumference stood as its most salient feature in the cityscape. In its massing of domes to support the central core, Mehmet's mosque shared a general plan and elevation with descriptions of the two principal churches of the earlier emperors in Constantinople, the seat of the Bishop, the Hagia Sophia and the Holy Apostles.[10] Mehmet's building campaign at the site of the Holy Apostles related to other interventions undertaken by the Sultan in the stratigraphy of imperial Greek sites. Soon after he assumed power, Mehmet II converted both the Hagia Sophia and the Parthenon, into mosques.[11] Though its accuracy cannot be confirmed, an inscription on the portal of the mosque reads that it was begun on February 21, 1463, around the time that Venice and its allies prepared a declaration of war against the Ottomans.[12] Given its significance as a locus of imperial identity, the Sultan's choice of the Holy

3. Exterior view of the domes of San Marco, Venice, twelfth century, covered with lead in the thirteenth century. Photograph by author.

Apostles for his mosque formed part of a larger claim to an inheritance of the city as a historical seat of power, and to a more decisive authority over the Byzantine past as the West prepared to challenge that claim.

An Ottoman declaration on the construction of churches in Constantinople demonstrates the nuanced attitudes toward assertions of Christian identity in the city. The policy states, "We shall not try to turn their churches into mosques, but they are not to build any new churches."[13] By silencing the bells of the churches—civic signs of autonomy—the Ottomans struck the sounds of Christianity from the aural landscape of the city. The prevention of further building was meant to decrease the number of monumental signs of Christianity in the city. Records of the conversion of Constantinople into an Ottoman city suggest that the Ottomans engaged in what Jas Elsner refers to as "the public gesture of forgetting" in his study of the preservation of monuments in ancient Rome subject to vandalism through *damnatio memoriae*.[14]

In her recent book on the transformation of Constantinople into an Ottoman city, Çiğdem Kafescioglu contends that the appropriation of the site of the Holy Apostles for Sultan Mehmet II's *külliye* was likely calculated as an act of aggression against Byzantine imperial identity. However, Kafescioglu emphasizes that several other factors also contributed to the destruction of the Church and the location of the Mosque.[15] The Church of the Holy Apostles may not have been demolished as an act of political hostility towards the Byzantine Empire, but it signaled transformation and reconstruction. In 1420 Cristoforo Buondelmonti

4. Façade of the Basilica of San Marco, Venice, begun in the thirteenth century. Photograph by author.

noted that the church was in a state of disrepair after years of neglect.[16] After assuming control over Constantinople, Mehmet bequeathed the Church of the Holy Apostles to the Greek Patriarchy, but they abandoned it soon after. Thus, when the builders of the Fatih Mosque incorporated columns of *verde antica* and granite from the Holy Apostles into the hospice of the mosque complex, they may have been taking advantage of the quarry made available by the compromised church.[17]

Despite the condition of the church by the end of the fifteenth century, Mehmet's courtiers colored the construction of his mosque in the hues of rivalry in descriptions of Mehmet's capture of Constantinople. Kritovoulos, a Greek working in the service of the Ottomans soon after 1453, wrote that "The Sultan himself selected the best site in the middle of the city and commanded them to build a Mosque which in height, beauty, and size should compete with the largest and finest temples already existing there."[18] An Ottoman ambassador, Tursun Beg, similarly noted that the Mosque was designed to surpass the monuments erected by the Byzantine emperors in Constantinople: "And he built a Great Mosque based on the design of Ayasofya, which not only encompassed all the arts of Ayasofya, but in addition, incorporated modern features constituting a fresh new idiom unequalled in beauty."[19] The spared Hagia Sophia remained as a technical paradigm for the Fatih Mosque complex with its larger dome, built on a prodigious scale in a more elevated area of the Constantinople to indicate Mehmet's

5. Hagia Sophia, Istanbul, fifth century. Photograph by author.

political, cultural, and religious ascendancy in the city that he conquered.

Tursun Beg described the Hagia Sophia in soaring tones, but his exposition also advances a broader theory of the structure as a meditation on the transience of the physical fabric of Constantinople. He declares that the mosque's builders surpassed Byzantine architecture by employing superior technical skills.

> What a dome, that vies in rank with the nine spheres of heaven! In this work a perfect master displayed the whole of the architectural science. Mehmet II, having looked upon the strange and wondrous images and adornments that were on the concave inner surface, deigned to climb up to the dome. When he saw the dependent buildings of this mighty structure fallen in ruin, he thought of the impermanence and instability of this world, and of its ultimate destruction.[20]

As tangible proof of his growing clout, Mehmet was able to transform the great and intricate monuments of the Byzantine Empire by reducing them to rubble. In extolling Mehmet through his buildings, Beg drew on Byzantine accounts that promoted the grandiosity of the churches built by the Eastern emperors. Both Constantine of Rhodes in the tenth century and Mesarites in about 1200 portray the Holy Apostles as having surpassed human building capabilities. In one passage, Mesarites described the domes of the church, "As if towering giants had come forth and extended their hands into the air, weaving together fingers."[21] These descriptions of an awe-inspiring building with an intricate, anthropomorphic, vaulting system may have served to restore dignity to the Holy Apostles after years of neglect since its patronage by the early emperors. That later Ottoman writers similarly promoted the technological feats of Byzantine building demonstrates the role of monuments in shaping the rhetoric of appropriation.

While many Ottoman court historians described Mehmet's takeover and destruction of significant Byzantine monuments to demonstrate his authority through conquest, Western merchants and envoys stationed in Constantinople recorded their impressions of the physical fabric of the city as a rally cry to war. Accounts of the conquest of Constantinople by Westerners often mourned the loss of Byzantine buildings as a synecdoche for the dismantling of Byzantine rule by the Ottoman Turks. One elegy, written by a Venetian, lamented the conversion of the Hagia Sophia into a mosque before he beseeched Justinian and his wife Theodora to witness what had become of the Holy Apostles. "A great church, the great cry, The Apostles. With a high voice, it says, O Empress Theodora, Come and See your sacred edifice. I do not believe I have ever thought it would come to this."[22] Passages like this, linking the city's ecclesiastical monuments to the beginning of the Eastern Empire and an embrace of Christianity, confirms an incipient belief in Susan Stewart's contention that "writing promises . . . the immortality of the material world."[23] Accounts of the destruction of Byzantine architecture by Mehmet's courtiers and Westerners alike, reify the infrastructure of the city as a representation rather than as a functional monument.

Many Western writers demonstrated familiarity with the Holy Apostles as a reliquary and the general significance of relics in Constantinople. In his autobiographical *Commentaries,* written before his death in 1464, Pope Pius II alluded to spoliation of the Holy Apostles in his imagined reaction to the conquest by Emperor Constantine, the founder of the city. He wrote, "I have been troubled to learn that the city you bade me to found, the most holy city of Christendom, second only to Rome, has been occupied by the satellites of Mehmet. The famous temples where your name was praised have been polluted . . . the sepulchers of my successors destroyed, the most holy relics have been thrown to the dogs."[24] Mehmet's admirer, Kritovoulos, similarly described "things they threw in dishonor on the ground—ikons and reliquaries and other objects from the churches."[25]

Testimonies about the desecration of Byzantine architecture as casualties of war coincided with Pope Pius II's papal proclamation in 1462 that prohibited the despoliation of ancient monuments for construction material in order that "our venerable city be preserved in its dignity and splendor."[26] The ruins of ancient Rome stood as a sign of decay, but the pope promoted the archaeological expeditions that were becoming popular in humanist circles as a form of ancient learning.[27] According to the pope, cohabitating with ancient "buildings permitted us to better perceive the fragility of human affairs," a sentiment that must have resonated with Europeans absorbing accounts of the fall of Constantinople and debating whether to rescue the physical remains of the East from the Ottomans.

As the Ottomans gained territory around the Mediterranean during the mid-fifteenth century, theologians, politicians, and exiles in the West debated an appropriate response to the conquest and the threat it posed to the preservation of symbolic monuments of Constantinople. Venetian merchants with trading interests in the East were not as eager to break diplomatic ties with the Ottoman Turks as others, like the Greek émigrés, who were anxious about the fate of their cultural patrimony. The Venetian Senate temporized when it came to dealing with the Sultan. In 1459, when Pius II announced his plans for a Crusade to reclaim the Holy Land and Constantinople, Venice prohibited the papal congress from meeting in its territory for fear that pro-Crusade sentiment would jeopardize Venetian business interests across the Adriatic.[28] Venice's entry into war and its eventual consent to the Crusade a year later in 1464 represented a notable change in policy after territory in the Aegean and along the Dalmatian coast had been conquered by the Ottomans.

The descriptions of ruins in Constantinople were complemented by the removal of sacred property to the West, including relics and manuscripts, with Venice serving as the principal beneficiary. As the Venetian senate debated a declaration of war on Mehmet, it ordered its soldiers to procure the head of St. George from Istanbul.[29] The acquisition of St. George's relics provided a conspicuous analogy to the current crisis in Venice, where in the fifteenth century, the West saw itself allegorically as a modern-day St. George, vanquishing the dangerous dragon as an emblem of the Ottoman Empire.[30] With the relics of George safely housed in Venice, the Republic could explicitly align the saint's struggle with its own against the Turks.

When the Fatih Mosque was under construction in the early 1460s, some of the relics of the principal saints associated with the Church of the Holy Apostles reemerged in the West. There are conflicting reports about the trajectory of many of the relics held at the Holy Apostles, particularly after the Venetians sacked the city during the Fourth Crusade of 1204. The Venetians returned to Europe with many precious items spoliated from Constantinople, then under Byzantine rule.

After 1204, the Amalfi Cathedral in southern Italy laid claim to the head of St. Andrew—considered the first patriarch of the Eastern Church and one of the principal saints displayed at the Church of the Holy Apostles—while pilgrims described seeing a part of the saint's head in the church of St. George of Mangana in Constantinople. However, the dominant story surrounding Andrew's remains is found in the autobiographical *Commentaries* by Pope Pius II, where he recounts the negotiations to translate the head of Andrew to Rome from the saint's birthplace of Peloponnesus soon after it fell to the Ottomans in 1460. Andrew's head was processed to the Pope on its arrival in Rome in a highly choreographed ceremony that emulated the Adventus ceremonies in Constantinople. The procession culminated in an anti-Turkish speech by the Greek Cardinal Bessarion, who proffered the relic to point out the vulnerability of other holy objects scattered throughout Ottoman-held territory.[31] The Cardinal told his listeners that St. Andrew's head itself had spoken and asked the Pope to "destroy the barbarians and defend the church," as his brother, St. Peter, would have wanted.[32] Because Andrew was considered the first emperor in the East, the translation of Andrew's relics symbolically unified the Latin and the Eastern Church as Bessarion made an appeal to the spiritual head of the Latin Church, the pope, by invoking his predecessor, Peter.

The account of Andrew's entrance into Rome coincided with an attempt to translate to Venice the relics believed to have belonged to St. Luke the Evangelist, another saint prominently displayed at the Holy Apostles.[33] In 1463, the leader of Venice, Doge Cristoforo Moro, arranged for the removal of an arm believed to be St. Luke the Evangelist to Venice from an area under the threat of Turkish control. While Mehmet appropriated the site of the Church of the Holy Apostles as an emblem of his authority over Byzantium, the Venetian doge began to embrace the receding Empire as a part of Venetian history. He installed what he advertised as the relics of St. Luke the Evangelist in the church of San Giobbe, a small church that he had been supporting at least since the 1450s, before he was Doge.

The Paduans mounted a challenge to the authenticity of the Venetians' claim to the torso of St. Luke the Evangelist for they laid claim to the same body part. In his decision, the emissary of the pope, Cardinal Bessarion, adjudicated in favor of Venice over the identification of the relic. Bessarion may have authenticated the relic of St. Luke in favor of Venice as a strategic move to secure the doge's commitment to participate in the Crusade against the Turks as they took over Bessarion's native Greece. However, the papacy quickly realized that the ruling in favor of Venice negated the identity of the attribution of a head in the Vatican to Luke the Evangelist, which was believed to have formed a perfect fit with the body in Padua. Despite his Venetian pedigree, Pope Paul II reversed Bessarion's decision, and the relic at San Giobbe was identified as belonging to St. Luke of Stiris.[34] The friars at San Giobbe, however, remained intransigent about the original attribution: in 1525, a pilgrim en route to the Holy Land had gone to San Giobbe, where he was shown "the entire body of St. Luke the Evangelist."[35]

As the Venetian Republic prepared to declare war on the Ottomans in 1463, the presence of yet another evangelist in Venice, this time Luke, would promote the idea that the city was a safe haven for the authors of the gospels and property sacred to Christians. Neither the relics of Andrew nor Luke that were imported onto the Italian peninsula could be securely linked to those relics that were once at the Holy Apostles, but the two forces allied against the Ottomans, the Venetian doge and the pope, both claimed to possess relics of saints who had been associated with the Church of the Holy Apostles as they prepared to crusade. As Patrick Geary and Thomas Dale have argued, relics were often promoted during the middle ages and the Renaissance as sacred objects that offered uncontested political legitimacy.[36] The relics served as indices of a Byzantine imperial power that was now contested by Ottomans.

The Venetians had political motive to preserve the identification of the relics as Luke the Evangelist, for the acquisition of an evangelist's relics recalled one of the foundational events of Venetian sovereignty: the narrative of the journey of St. Mark's relics from Alexandria to the lagoon in the ninth century.[37] When Mark's relics arrived from Alexandria, Venice was under the control of the Byzantine Empire, but it was in the process of instituting an elected leader, the doge, to establish itself as an independent power. The bishop led the relics of Mark in a procession to San Teodoro, the private chapel of the doge, who vowed to build a reliquary church for St. Mark nearby. When he installed St. Luke's relics at San Giobbe in the fifteenth century, it seems that Doge Moro tried to create a similar alliance between the small church that he patronized on the outskirts of Venice and San Marco. The half-length reliefs of the apostles that line the intrados of the high altar chapel of San Giobbe point to the importance placed on the followers of Christ at the church.

Stylistically, there are fundamental similarities between the San Marco and San Giobbe. San Giobbe is covered by a low dome on pendentives, pierced by eight deep windows, like the dome of San Marco. Given that the patron of San Giobbe, Moro, was a doge, who had just choreographed the translation of St. Luke the Evangelist's relics to the church, the dome was likely conceived to create an alliance with San Marco as an

6. Dome over the High Altar Chapel, San Giobbe, Venice, fifteenth century. Photograph by author.

evangelist's reliquary and as the basilica that hosted the doges in their official capacities. By drawing on the design motifs of San Marco, the dome profile of San Giobbe also coopted an architectural and iconographical vocabulary associated with the Byzantine Empire. Though small, the roundel of God the Father that marks the summit of the dome of San Giobbe evokes descriptions of the mosaic of the Pantocrator in the Holy Apostles. It suggests that the designers of San Giobbe drew not only on Byzantine decorative convention but ekphrases that posited domes as metaphors of the heavenly realm.

The motif of the low dome on pendentives appeared in a spate of small centrally planned churches built in Venice during the late fifteenth century. A few of these churches, including San Giovanni Crisostomo and Santa Maria Formosa, were heavily reconstructed in the late fifteenth century after they had been destroyed by fires in the mid-1470s.[38] The architect of both churches, Mauro Codussi, may have based his designs on

6a. Dome over the High Altar Chapel, San Giobbe, Venice, fifteenth century. Photograph by author.

preexisting plans, but as James Ackerman argued, the presence of these churches designed on a square base of nine bays broke with the mendicant convention of building churches with elongated naves that had been taking hold in Venice, pointing instead to Eastern convention.[39] Many of these plans recall the plan of San Marco and the Holy Apostles, as reconstructed by archaeologists. In December 1499, a "madonna Grecha" was commissioned for San Giovanni Crisostomo.[40] None of the supporters or patrons of the church were known to have personally traveled to the East, but it seems that in their designs they were promoting the notion of a "Greek" style as Byzantine culture in the East was perceived to be under threat.

Though these smaller Venetian churches shared design motifs with the basilica of San Marco, their appearance soon after reports of the desecration of the Constantinople circulated in Venice can also be attributed to the rhetoric that was redefining the Republic's understanding of its relationship to Byzantium. When the Venetians sacked Constantinople in 1204, they famously returned to Venice with spoils from

7. San Marco, central dome, Venice, ninth century, rebuilt in the twelfth century. Photograph by author.

sites of Byzantine power, which they displayed at the basilica San Marco, the city's nexus for civic and religious ceremony. Because many of the objects spoliated were used to decorate San Marco, the materials became signifiers of Venetian cultural heritage, linking Venice's victory to the claim of St. Mark's relics. Sacred objects that once shaped the authority of the Byzantine Empire were stolen by the Venetians as a sign of hegemony and affixed onto the surfaces of San Marco. In contrast to the Venetian spoliation of Constantinople in 1204, relics and architectural forms appropriated as salvage from the Turks were propped up as expressions of a Byzantine Golden Age. Sacred objects and architectural forms associated with the East promoted anti-Turkish sentiment, eclipsing the more nuanced political and economic tensions that had defined the relationship between the former Constantinople and the Venetian Republic.

In 1468, Cardinal Bessarion donated several hundred of his Greek and Latin manuscripts to the Republic from his endangered homeland and a reliquary containing a relic of the True Cross, to Venice.[41] The relic had been given to Bessarion as a gift by Gregory, the patriarch of Constantinople, when he died in Italy in 1459. In a letter to the doge explaining his regifting of the items to Venice, Bessarion referred to the Republic as "another Byzantium," suggesting that Venice was a worthy inheritor of his cultural patrimony.[42] An inscription on Bessarion's relic of the True Cross mentions Irene, niece of the Byzantine emperor, initiating an imperial chain back to Constantine and

8. Interior, San Giovanni Crisostomo, Venice, rebuilt late fifteenth century. Photograph courtesy Didier Descouens, Wikimedia Commons

his mother, Helena, who brought the relic to her son's new city. As Byzantium was threatened, scholars transferred to Venice their praise of the cultural and spiritual golden age that they believed had existed in Constantinople. Venice became the site where the rich cultural heritage of Byzantium could be memorialized, resurrected, and celebrated. The translation of objects and architectural forms to the West aided in the construction of a more harmonious and seamless history between Venice and Byzantium, glossing over the divisions that had actually defined the relationship between the two powers for almost a thousand years.

Biography

Janna Israel is assistant professor of art history at Virginia Commonwealth University. She received her PhD from the Massachusetts Institute of Technology. She is completing a book on Venetian identity after the conquest of Constantinople by the Ottoman Turks. Her research has been supported by the American Academy in Rome and the Delmas Foundation. Prior to arriving at VCU, Professor Israel was a postdoctoral research associate at the Center for Advanced Study in the Visual Arts at the National Gallery of Art.

Notes

1 Cyril A. Mango, *Le développement urbain de Constantinople, IVe–VIIe siècles* (Paris: De Boccard, 2004), 27. W. Kleiss, *Topographisch-archaologischer Plan von Istanbul* (Tubingen: E. Wasmuth, 1965), 8 n.35; Gülru Necipoğlu, *Architecture, Ceremonial, and Power: The Topkapi Palace in the Fifteenth and Sixteenth Centuries* (Cambridge, Mass.: MIT Press, 1991), 14–15; Marcell Restle, "Filarete in Konstantinopel," *Pantheon* 39 (1981): 361–67; Mehmet Aga-Oglu, "The Fatih Mosque at Constantinople," *Art Bulletin* 12 (1930): 179–95.

2 Sabine MacCormack, "Change and Continuity in Late Antiquity: The Ceremony of 'Adventus,'" *Historia: Zeitschrift für Alte Geschichte* 21 (1972): 721–52; in his sixth century *De Aedificiis,* Procopius wrote that Constantine's son "built this church in honor of the Apostles . . . decreeing that tombs for himself and for all future Emperors should be placed there." Ann Wharton Epstein, "The Rebuilding and Redecoration of the Holy Apostles in Constantinople: A Reconsideration," *Greek,*

Roman, and Byzantine Studies 23 (1982): 79–92; Glanville Downey, "The Tombs of the Byzantine Emperors at the Church of the Holy Apostles in Constantinople," *The Journal of Hellenic Studies* 79 (1959): 27–51. The *Book of Ceremonies,* written in the tenth century, provides a lengthy catalogue of imperial burials at the Church.

3 Pierre Nora, *Realms of Memory: Rethinking the French Past,* trans. Arthur Goldhammer (New York: Columbia University Press, 1996–ca. 1998), xvii.

4 Glanville Downey, "Nikolaos Mesarites: Description of the Church of the Holy Apostles at Constantinople," *Transactions of the American Philosophical Society* (1957): 271.

5 Richard Krautheimer, "Justinian's Church of the Holy Apostles," *Studies in Early Christian, Medieval, and Renaissance Art* (New York: New York University Press, 1969). Krautheimer identified four images of the Holy Apostles: three in the Menologium of Basil II (Vat. Gr. 1613) and one from the sermons of James Kokkinobaphos (Vat. Gr. 1162).

6 Otto Demus, *The Church of San Marco in Venice: History, Architecture, Sculpture, with a Contribution by Ferdinando Forlati* (Washington, D.C.: Dumbarton Oaks Research Library and Collection, 1960), 44. Deborah Howard, *The Architectural History of Venice* (New Haven, Conn.: Yale University Press, 2002), 19–22.

7 Richard Krautheimer, *Early Christian and Byzantine Architecture* (New York: Penguin Books, 1986), 69–70; Cyril Mango, *Byzantine Architecture* (New York: H. N. Abrams, 1976), 296. Procopius, *De Aedificiis,* trans. W. Lethabv and H. Swainson (New York: 1894), 24–28, books I–IV.

8 Juergen Schulz, "La Piazza Medievale di San Marco," *Annali di Architettura* 4/5 (1992–93): 134–56.

9 "Il paradigma della memoria: San Marco a Venezia e la chiesa dei Santi," *Medioevo: Immagine e memoria* (2009): 127–31.

10 Gülru Necipoğlu, *The Age of Sinan: Architectural Culture in the Ottoman Empire* (London: Redaktion, 2005), 84–86. Necipoğlu has identified a fragmented plan on Italian paper from the second half of the fifteenth century as Mehmet's *külliye,* due to the depiction of a forecourt with domical arcades that are characteristic features of most sultans' mosques.

11 Robert Ousterhout, "Bestride the Very Peak of Heaven: The Parthenon after Antiquity," in *The Parthenon: From Antiquity to the Present,* ed. Jenifer Neils (Cambridge: Cambridge University Press, 2005), 317–18.

12 H. Edhem, *Nos mosquies de Stamboul* (Istanbul: n.p., 1934), 48.

13 Treaty with the Genovese Merchants in Constantinople.

14 Jas Elsner, "Iconoclasm and the Preservation of Memory," in *Monuments and Memory, Made and Unmade,* eds. Robert S. Nelson and Margaret Olin, 209–31 (Chicago: University of Chicago Press, 2003). See also Annabel Wharton, who evaluated the destruction of sites in the history of late ancient Judaism through Jacques Derrida's conception of erasure. In this conception, erasure is a "complex performance" that "establishes the eraser's authority to suppress. But erasure leaves a mark that frustrates the gesture and reverses its effect." Annabel Wharton, "Erasure: Eliminating the Space of Late Ancient Judaism," *From Dura to Sepphoris: Studies in Jewish Art and Society in Late Antiquity,* eds. Lee I. Levine and Zeev Weiss, 195–214 (Portsmouth, R.I.: Journal of Roman Archaeology, 2000). For more on the vandalism of monuments as acts of political aggression, see Dario Gamboni, *The Destruction of Art: Iconoclasm and Vandalism since the French Revolution* (London: Reaktion Books, 1997), 32; Erika Naginski, *Sculpture and Enlightenment* (Los Angeles: Getty Research Institute, 2009).

15 Çiğdem Kafescioglu, *Constantinopolis/Istanbul* (University Park: Pennsylvania University Press, 2010), 5–10.

16 George Sphrantzes, "Chronicon Minus," in *The Fall of the Byzantine Empire,* trans. Marios Philippides, 99–151 (Amherst: University of Massachusetts Press, 1980); *The Mosaics and Frescoes of St. Mary Pammakaristos (Fethiye Camii) at Istanbul,* eds. Hans Belting, Cyril A. Mango, and Doula Mouriki (Washington DC: Dumbarton Oaks, 1978), 26. Cristoforo Buondelmonti, *"Descriptio urbis Constantinopoleos," Corpus Scriptoum Historiae Byzantinae* (Bonn, Switzerland: n.p., 1836), 181.44.

17 Pierre Gilles, *The Antiquities of Constantinople,* trans. John Ball (New York: Italica Press, 1988), 222. In the sixteenth century, Peter Gilles noted the materials in a courtyard of the hospice near the mosque. Bolstering the accuracy of Gilles's description, archeologists have recently unearthed what they argue are the Church's walls near the charitable kitchen, located southeast of the mosque. Ken Dark and Özgümüş Ferudun, "New Evidence for the Byzantine Church of the Holy Apostles from Fatih Camii, Istanbul," *Oxford Journal of Archaeology* 21, no. 4 (November 2002): 393–413. Albrecht Berger, "Streets and Public Spaces in Constantinople," *Dumbarton Oaks Papers* 54 (2000): 161–72.

18 Kritovoulos, *History of Mehmed the Conqueror,* trans. Charles T. Riggs (Princeton, N.J.: Princeton University Press, 1954), 140.

[19] Tursun Beg, *History of Mehmet the Conqueror,* trans. Halil Inalcik and Rhoads Murphey (Minneapolis: Bibliotheca Islamica, 1978), 7.
[20] Ibid., 7.
[21] Downey, "Nikolaos Mesarites," 271.
[22] Anonymous, "Lament on the Fall of Constantinople," in *La Caduta di Costantinopoli,* ed. Agostino Pertusi, 306–8 (Milan: A. Mondadori, 1976).
[23] Susan Stewart, *On Longing: Narratives of the Miniature, the Gigantic, the Souvenir, the Collection* (Baltimore: The Johns Hopkins University Press, 1984), 31.
[24] James Hankins, *Humanism and Platonism in the Italian Renaissance* (Rome: Edizioni di storia e letteratura, 2003–4). See also Lauro Quirini in Pertusi, *Caduta,* 52–54: "for not only has a royal city been captured, temples devastated and holy places polluted, but an entire race has been overcome—the name of Greece is blotted out. Over a hundred and twenty thousand volumes were destroyed, as I learn from Cardinal Isidore of Kiev."
[25] Kritovlous, *Mehmed,* 60–75.
[26] Cesare Donofrio, *Visitiamo Roma nel Quattrocento* (Rome: Romano società editrice, 1989), 25. Pope Pius II, Bull of April 28, 1462, "Cum almam nostram urbem in sua dignitate et splendore conservari cupiamus." Francoise Choay, *The Invention of the Historic Monument, trans. Lauren M. O'Connell* (Cambridge: Cambridge University Press, 2001), 34–35.
[27] David Karmon, *The Ruin of the Eternal City: Antiquity and Preservation in Renaissance Rome* (Oxford: Oxford University Press, 2011), 69–72.
[28] Giovanni Battista Picotti, *La Dieta di Mantova e la Politica de Veneziani,* ed. Gian Maria Varanini (Trent: Università degli Studi di Trento, 1996). Pope Pius II, *Commentaries,* vol. 1, ed. Margaret Meserve and Marcello Simonetta (Cambridge, Mass.: Harvard University Press, 2003), 205; Norman Housley, *Crusading in the Fifteenth Century: Message and Impact* (New York: Palgrave Macmillan, 2004); Jacques Paviot, *Les Ducs de Bourgogne, la Croisade et l'Orient* (Paris: Presses de l'Université de Paris-Sorbonne, 2003). James Hankins, "Renaissance Crusaders: Humanist Crusade Literature in the Age of Mehmed II," *Dumbarton Oaks Papers* 49 (1995): 111–207.
[29] Kenneth M. Setton, "Saint George's Head," *Speculum* 48 (1973): 9. In August, 1462, the head of St. George was found on the island of Aegina and transported to Venice. See also the *Acta Sanctorum,* Aprilis, Tom. III, 133.
[30] Stefano Carboni, *Venice and the Islamic World* (New Haven, Conn.: Yale University Press, 2007), 132.
[31] Pius II, *Commentaries,* book 8.
[32] Ibid.
[33] Biblioteca Museo del Correr, PDD, 727, vol. 1, 104.
[34] Archivio di Stato, Venezia (ASV), Procuratore di San Marco 84, Proc 189a, Fasc. 5. The 1634 inventory of San Giobbe notes that the church contains St. Luke the Evangelist's body.
[35] Iain Fenlon, "St. Mark's before Willaert," *Early Music* 21, no. 4 (November 1993): 547–63.
[36] Patrick Geary, *Furta Sacra: Thefts of the Relics in the Central Middle Ages* (Princeton, N.J.: Princeton University Press, 1978), 9–15; Thomas E. A. Dale, "Inventing a Sacred Past: Pictorial Narratives of St. Mark the Evangelist in Aquileia and Venice, ca. 1000–1300," *Dumbarton Oaks Papers* 48 (1994): 54, 100. For information on relics in Constantinople, see also Martina Bagnoli et al., eds., *Treasures of Heaven: Saints, Relics, and Devotion in Medieval Europe* (New Haven, Conn.: Yale University Press, 2010), 13.
[37] On the translation of Mark's relics, see Dale, "Inventing a Sacred Past," 54, 100. Geary, *Furta Sacra,* 88–94; Otto Demus, *The Mosaics of San Marco* (Washington, D.C.: Dumbarton Oaks Research Library, 1960), 45.
[38] Richard J. Goy, *Building Renaissance Venice: Patrons, Architects, and Builders, c. 1430–1500* (New Haven, Conn.: Yale University Press, 2006), 250–52.
[39] James Ackerman, "Observations on Renaissance Church Planning in Venice and Florence," in *Florence and Venice: Comparisons and Relations: Acts of Two Conferences at Villa I Tatti in 1976–1977,* 2 vols., eds. S. Bertelli, N. Rubinstein, and C.H. Smyth, 2:287–307 (Florence: La Nuova Italia, 1979). Olivato and Lionello Puppi, *Mauro Codussi* (Milan: Electa Editrice, 1977), catalogue nos. 9 and 11.
[40] ASV, Valverde, b. 66, fols. 157–58.
[41] Caroline Campbell, "The Bellini, Bessarion, and Byzantium," in *Bellini and the East,* eds. Caroline Campbell and Alan Chong (Boston: Isabella Stewart Gardner Museum, 2005), 38. The Reliquary of the True Cross arrived in Venice after Doge Moro's death in 1472.
[42] Biblioteca Nazionale Marciana, Lat.XIV (=4235) on Bessarion's donation of manuscripts. Lotte Labowsky, *Bessarion's Library and the Biblioteca Marciana: Six Early Inventories* (Rome: Edizoni di Storia e Letteratura, 1979), 24–25.

Book Review
George Wheeler

Conservation Takes a Reflective Turn

A Laboratory for Art: Harvard's Fogg Museum and the Emergence of Conservation in America, 1900–1950

Francesca Bewer
Cambridge, Mass.: Harvard Art Museum and Yale University Press, 2010

Francesca Bewer's new book, published jointly in 2010 by Harvard Art Museum and Yale University Press, takes a step in a new direction for conservation literature. *A Laboratory for Art: Harvard's Fogg Museum and the Emergence of Conservation in America, 1900–1950* traces the development of the Conservation Department in the Fogg Museum from its beginnings to just after World War II. That history is told largely by following the personal and professional growth of one man, Edward Waldo Forbes, and those he gathered around him during his tenure as the director of the Fogg Museum from 1909 to 1944. As Bewer (pronounced BAY·ver) points out, Forbes came from good stock—a Boston Brahmin and the grandson of the Transcendentalist poet and philosopher Ralph Waldo Emerson—and his grounding in the American Enlightenment played a large part in shaping both his character and ultimately that of the Fogg's Conservation Department.

The book contains a preface and five chapters, each chapter dealing with approximately a decade of the life of the Conservation Department. The preface is not to be overlooked, as it establishes two important themes for the book, and, though often undervalued or unstated, guiding principles for the practice of conservation: "Whether works of art are transformed by nature or by human intervention, they never endure in their original state," and "The care of artworks—their cleaning, repair, and preservation—is inevitably an act of interpretation, influenced by philosophical, historical, artistic, and scientific ideas as well as the tastes of a particular time and place."

"Forbes Takes on Flaking Paint (1899–1910)," chapter 1, first discusses Forbes's apparent epiphany when he wished to purchase a painting that was in poor condition. He quickly came to learn that there was little consensus on approaches to conservation to damaged works of art. Much was left to the subjective domain of the restorer, and, as a young man steeped in the Enlightenment confidence in science, he sought to bring its "objectivity" to bear on art conservation. The late nineteenth century was also a time Americans began to take a strong interest in collecting art, but many, like Forbes himself, lacked the experience and sophistication of European collectors. Feeling the discomfort of the uninitiated, Forbes also thought that a greater understanding of the materials and methods of art objects through science would assist in their authentication, heretofore left to connoisseurs like his highly regarded Cambridge compatriot Bernard Berenson.

In chapter 2,"The Museum as Laboratory (1910s)," Bewer outlines the background for Forbes's model of conservation—studio art or understanding of craft, experimental science, and art history—later dubbed the three-legged stool of conservation education by George Stout (see below). Forbes did not leave theory behind in establishing the Fogg's approach to conservation, as he was well aware of the conflict between Ruskin and Viollet-le-Duc and the attempted compromise offered by Camillo Boito. What Forbes stressed, and why he felt so strongly that these activities should take place in the university, is that transparency of thought and action in conservation was of the utmost importance for both the general public and the specialist. This transparency was in sharp contrast to the more secretive world of restorers of his time.

The next chapter, "Good Chemistry: Attracting Like Minds (1920s)," shows how Forbes brought together the core team of the Fogg that would be the heart and soul of the Conservation Department for nearly thirty years. Forbes first obtained the services of Daniel Thompson, trained as an undergraduate in chemistry at Harvard, who would apply that knowledge to medieval painting techniques and later pen *The Materials and Techniques of Medieval Painting*. Soon after, Alan Burroughs joined the Fogg staff. Continuing Forbes's desire to apply science to art, Burroughs specialized in the examination of paintings and objects using X-rays to reveal underlying layers or structural features. While Thompson and Burroughs played key roles in the development of the Fogg as a center for the technical study of art and the development of a more scientific foundation for conservation, it was the addition of George Stout and Rutherford Gettens that finally put the institution on the map. Stout had no science background but had developed a keen interest in the processes of making art through his own studies in drawing and painting. Conversely, Gettens, a scientist trained in inorganic chemistry at Harvard, "had never set foot in an art museum before Stout showed him around the Fogg." But a bond would form between them that would lead to their joint effort, *Painting Materials: A Short Encyclopedia,* first published in 1943, and, together with Forbes, they would turn the Fogg into "the leading center for conservation research and training in the United States" of its time.

Chapter 4, "A Department for Conservation and Technical Research (1930s)," demonstrates the growing international influence the center came to have, particularly through the direct training of the next generation of conservation and museum professionals and through the encouragement and advising to other institutions in establishing their own training centers. Key elements in the training program were the importance of documentation and the need for transparency and dissemination of information about the technical features and conservation methods applied to art objects through publication and attending meetings. Despite the heavy workload for the small staff, they still managed to launch and keep alive the Fogg's journal, *Technical Studies in the Field of the Fine Arts,* the first of its kind, published from 1932 to 1942.

The final chapter, "War as a Catalyst for Conservation (1940s)," shows that the Fogg's Conservation Department continued to have a strong influence in conservation activities worldwide and was instrumental in establishing standards and practices for the protection, moving, and storage of works of art during wartime. As the war neared its end, Forbes and his long time codirector, Paul Sachs, retired and the direction of the Fogg's Conservation Department began to drift. Soon after, Stout became director of the Worcester Art Museum and Gettens went off to Los Alamos for a year. Upon returning, Gettens teamed with Richard Buck, now head conservator (and a Fogg trainee), to keep the institution going, but by 1951 the Conservation Department was a shadow of itself. That year, Gettens left for the Freer Gallery of Art, Buck to Ohio where he established the Intermuseum Conservation Center, and others of the conservation team to important institutions or positions. However, even as the Fogg's main players departed, they took with them the knowledge they developed, the lessons they learned, and an abiding passion for the field of conservation and the desire to extend its influence and secure its integrity. Extending the Fogg's principles of the professional grounding of conservation and transparency and dissemination of information about conservation, Stout and Gettens, along with American and European colleagues, established the International Institute for Conservation and the associated journal, *Studies in Conservation,* in 1950. And it would only be a small stretch to say it all began with "Forbes takes on flaking paint."

So how does Bewer's book take a "step in a new direction" for the conservation literature? Bewer essentially provides a history of the development of art conservation in America that leads to the creation of the university-based training programs. What she provides, however, is a *critical* history, not just a narrative history. Unlike many of those narrative histories, which range somewhere between puff pieces for contemporaries and hagiographies for the founders of the profession in America, Bewer does not shy away from controversy. She makes it clear that Forbes and his colleagues were not above criticism for their “scientific conservation.” While this moniker may not be accurate and Forbes’s confidence in the objectivity of science is harder to justify today than in his own time, it nonetheless conveys the sense of an approach to conservation sometimes pejoratively characterized as “letting the white coats into the conservation studio.” As pointed out by Bewer, many see this approach leading to a “hygienic interest” in an art object in contrast or opposition to an “aesthetic interest.” The dynamic and elastic tension between these interests persists to this day, with the recognition of the additional influences of historical and social values that attach to art objects. Some have gone even further to suggest that the reliance on science and the scientific approach to conservation leads to overzealous and overconfident interventions, and, consequently, aesthetic and historical conservation “disasters,” such as the cleaning of the Sistine Chapel ceiling as viewed by James Beck. In this vein, Bewer does not become squeamish when discussing Thompson’s horrible results (and questionable ethics) in removing a series of wall paintings from China in the 1920s and 1930s.

In a field that tends to shy away from or fully resist both external and self-criticism or even self-reflection, particularly in the United States, Bewer’s clear-eyed view is refreshing. In contrast, it was quite disturbing to witness the hostile reception Beck received at the American Institute for Conservation Annual Meeting in Portland, Oregon in 2004. Beck, who died in 2007, was highly critical of the field of conservation in general. At times his rhetoric and mode of presentation undermined, obscured, or even obliterated his message. However, his call for transparency and engagement with allied professions was exactly what Forbes and his colleagues at the Fogg fought for and what conservators are at least ostensibly trained to do. Bewer’s more open-minded and critical approach follows and reinforces a more recent trend in the conservation literature perhaps begun with the Getty’s publication of *Historical and Philosophical Issues in the Conservation of Cultural Heritage* in 1996. It took nearly ten years for this publication to bear fruit, and 2005 witnessed the publication of the first full-length English language edition of Cesare Brandi’s classic text *Theory of Restoration,* translated by Cynthia Rockwell. That same year, Salvador Muñoz Viñas published his *Contemporary Theory of Conservation* (the first theory book written by a conservator in modern times), in which he portrayed Brandi’s volume somewhat like “old testament” conservation theory. Muñoz Viñas ably showed the erosion of objectivity and injection of cultural relativity in the twentieth century and the impact of those trends on conservation. Both Brandi and Muñoz Viñas ask conservators to be more thoughtful, thoughtfully creative (for a time creativity was banned from conservation education), and more open to both outside influence and to discussing internal conflict. In 2009, Paul Eggert's *Securing the Past* and *Conservation: Principles, Dilemmas, and Uncomfortable Truths*, edited by Alison Richmond and Alison Bracker, each pushed the conservation community toward other disciplines and taking in hand those disciplines' tools of criticism. Like Beck, the two books emphasize that conservation has often been rather isolated and has not taken advantage of what can be learned from other critical approaches. Coming out of that isolationist view and exploring multiple avenues of criticism requires a full ingestion and metabolizing of Bewer's fundamental point that, "The care of artworks — their cleaning, repair, and preservation — is inevitably an act of interpretation, influenced by philosophical, historical, artistic, and scientific ideas as well as the tastes of a particular time and place." What *is* fundamentally different from other disciplines is that conservators work on actual physical objects and that their acts of criticism are often undoable. Brandi states this succinctly and profoundly, "Only the material form of the work is restored" (not the spirit, not the soul, and not the unknowable original intent). The conservator's act on the material or the material form is concrete and yet the frame of operation and the means for measuring success are subjective. The tension between the concrete act and the subjectivity surrounding the evaluation of the outcome often stays the hand or is a source of anxiety for a conservator. It is perhaps this nerve that Beck touched in his trenchant criticism of the field of conservation.

A few words should be said about the production of the book, which is offered only in paperback and at the modest price of $40. Contrasting the serious subject matter and matter-of-fact writing style (quite suited to the subject) is the playful use of brightly colored pages and sections of the book: powder blue for the preface, bright orange separating chapters, acid green for the notes, and bright yellow for the bibliography and index. Physically, the book is truly "handy" at 6 x 9 inches and a little over an inch thick, but it does not wear well, as the front and back covers appear to be glued flat against the spine from which they easily pull away.

These minor limitations aside, Bewer's book demonstrates the elemental role the Fogg played in training museum professionals for preservation and conservation, providing a model for conservation laboratories in American museum settings (and how Forbes both adapted European models for this as well as rejected some of their principles), and establishing protocols for the scientific examination of works of art. In addition, it is a model of thoughtful and reflective examination of an important part of the development of conservation in America and a sign of growth in the discipline itself.

Biography

George Wheeler joined the Historic Preservation Program of Columbia University in 2005 as director of conservation, after twenty-five years at the Metropolitan Museum of Art as a research scientist in the Sherman Fairchild Center for Objects Conservation, a post he continues to hold part-time. Probably best known for his research and writing in stone conservation, he also has taught in the graduate program for art conservation at New York University's Conservation Center for more than fifteen years and has consulted in architecture conservation since 1980.

1. Seizure, 2008. An Artangel / Jerwood Commission. Harper Road, London. Courtesy Corvi-Mora, London.

2. Copper Sulphate Chartres, Copper Sulphate Notre-Dame, 1997. Copper sulphate, Perspex, wood, card. 137 x 125 x 65 cm, RH97-1. Courtesy Corvi-Mora, London.

3. The Architect's Mother, 2003. Copper sulphate, BMW engine, steel, card. 230 x 128 x 86 cm, RHi03-9.
Courtesy Corvi-Mora, London.

Submission Guidelines

Future Anterior is a peer-reviewed (refereed) journal that approaches the field of historic preservation from a position of critical inquiry. A comparatively recent field of professional study, preservation often escapes direct academic challenges of its motives, goals, forms of practice, and results. *Future Anterior* seeks contributions that ask these difficult questions from philosophical, theoretical, and critical perspectives.

Articles should be 4,000 words, with up to seven illustrations. It is the responsibility of the author to secure permissions for image use and pay any reproduction fees. An abstract (200 words), a brief author biography (around 100 words), and a list of numbered image captions with credits must accompany the text. Acceptance or rejection of submissions is at the discretion of the editorial staff. Please do not send original materials, as submissions will not be returned.

Formatting requirements for the manuscript: Text must be formatted in accordance with the *Chicago Manual of Style,* 16th Edition. All articles must be submitted in English, and spelling should follow American convention. All submissions must be submitted electronically, on a CD or disk, accompanied by hard copies of text and images. Text should be saved in Microsoft Word or RTF format.

Formatting requirements for illustrations: Images should be sent as TIFF files with a resolution of at least 300dpi at an 8"-by-9" print size. Each image file should be numbered in accordance to the image captions. Figure placement should be indicated clearly in the text, after the paragraph in which they are referenced. Image captions and credits must be included with submissions.

Checklist of documents required for submission:

__ Abstract (200 words)
__ Manuscript (4000 words)
__ Illustrations (7)
__ Captions for illustrations
__ Illustration copyright information
__ Author biography (100 words)

Please mail all submissions to:
Future Anterior
Historic Preservation Program
Graduate School of Architecture, Planning, and Preservation (GSAPP)
400 Avery Hall
1172 Amsterdam Avenue
Columbia University
New York, NY 10027

Questions about submissions or published articles can be mailed to the above address or to futureanterior@columbia.edu

www.arch.columbia.edu/futureanterior

Become a *Future Anterior* Sponsor

Future Anterior is funded by grants and distributed in the spirit of making knowledge available to everyone. Donations are critical to helping us accomplish this mission. If you would like to become a sponsor, please fill out this page and mail it—with a check payable to "Columbia University"—to:

Future Anterior
Graduate School of Architecture, Planning, and Preservation
400 Avery Hall
1172 Amsterdam Avenue
Columbia University
New York, NY 10027

Or contribute by credit card online at:
http://www.arch.columbia.edu/giveonline/GSAPP.

All sponsors will be recognized in each issue of *Future Anterior* and receive a one-year subscription to the journal.

Please check the appropriate sponsorship level:

Individual Sponsor:	❑ begins at $100/year
Institutional Sponsor:	❑ begins at $500/year
Patron:	❑ begins at $1000/year

Name*: ______________________________

Address: ______________________________

____

City: ______________________________

State: ______________ Zip: ______________

Country: ______________________________

Institution/Office: ______________________________

E-mail: ______________________________

*please provide your name exactly as you would like it to appear in print